Yoshio Nishina
Father of Modern Physics in Japan

Yoshio Nishina

Father of Modern Physics in Japan

Dong-Won Kim

Taylor & Francis
Taylor & Francis Group
New York London

Taylor & Francis is an imprint of the
Taylor & Francis Group, an **informa** business

CRC Press
Taylor & Francis Group
6000 Broken Sound Parkway NW, Suite 300
Boca Raton, FL 33487-2742

International Standard Book Number-13: 978-0-7503-0755-0 (Hardcover)

Library of Congress Cataloging-in-Publication Data

Kim, Dong-Won, 1960-
Yoshio Nishina : father of modern physics in Japan / by Dong-Won Kim.
p. cm.
Includes bibliographical references and index.
ISBN 978-0-7503-0755-0
1. Nishina, Yoshio, 1890-1951. 2. Physicists--Japan--Biography. I. Title.
QC16.N56K56 2007

530.092--dc22
[B] 2007000832

Visit the Taylor & Francis Web site at
http://www.taylorandfrancis.com

and the CRC Press Web site at
http://www.crcpress.com

Preface

Yoshio Nishina has been one of my hero scientists ever since the mid-1990s when I wrote my first paper on the history of Japanese physics. I realized at once that he had made great contributions to the development of physics in Japan in the early twentieth century. I also discovered that there were few scholarly studies on him, and decided to write a short paper in order to analyze his role in the Japanese physics community in the 1930s and 1940s.

In March 1999, I interviewed Professor Emeritus Yoichiro Nambu of the University of Chicago with Professor Emeritus Laurie M. Brown of Northwestern University. Both Nambu and Brown were surprised when I naively asked questions like “How good a physicist was Nishina?” That interview not only reconfirmed the importance of Nishina in the history of physics in Japan but also led me to abandon my original plan for a paper in favor of writing a full biography of Nishina. After returning from Chicago, I received an e-mail from Brown. He stated that both Nambu and he agreed that I should write a biography of Nishina. Brown even offered to arrange for it to be published by the Institute of Physics (IOP) in Bristol, England. I hesitated for a while but finally decided to take on this great intellectual challenge. The Yoshio Nishina project began in the summer of 1999.

I would like to add special notes for Japanese titles and the arrangement of Japanese names in the book. Although family names precede given names in Japan, I have put given names first for two reasons. First, those Japanese scientists who are featured in this book published their most important works in English or in German, and adopted the Western way of presenting their names in their papers. So, the central figure of this biography was known to the scientific world as “Yoshio Nishina,” not as “Nishina Yoshio.” Second, this book is intended for the English speaking readers who are not familiar with the East Asian naming convention. I translated most of Japanese titles of books and papers into English and added “(Japanese)” at the end, but simply romanized the Japanese titles of journals or publishers.

Dong-Won Kim

Author

Dong-Won Kim was born in Seoul, South Korea, in 1960. He received his doctoral degree from Harvard University in 1991. His thesis was on the history of the Cavendish Laboratory at the University of Cambridge. He has taught history of science and technology at several universities in South Korea and the United States, including Korea Advanced Institute of Science and Technology (KAIST), Seoul National University and Johns Hopkins University. He is now visiting associate professor at Johns Hopkins University.

The list of his publication includes *Leadership and Creativity: An Early History of the Cavendish Laboratory, 1871–1919*, "J.J. Thomson and the Emergence of the Cavendish School, 1885–1900", "The Emergence of Theoretical Physics in Japan: Japanese Physics Community between the Two World Wars", "Winning Markets or Winning Nobel Prizes?: KAIST and the Challenge of Late Industrialization" (with Stuart W. Leslie), "Two Chemists in Two Koreas", and "Yoshio Nishina and Two Cyclotrons". His next book is on the history of science and technology in South Korea in the last half of the twentieth century.

Acknowledgments

During the project, I have been very lucky to receive a great deal of support from many people. First of all, I would like to express my deepest gratitude to Yoichiro Nambu and Laurie M. Brown who introduced me to this wonderful opportunity. I also would like to thank Silvan S. Schweber, Ewrin N. Hiebert, Shigeru Nakayama, and Takehiko Hashimoto for their encouragement and interest. Sam Schweber read the final draft and gave me a lot of worthwhile criticism and comments for revision. A condensed version of Chapter 6 of the book was published in the *Historical Studies in the Physical and Biological Sciences*, and I am grateful to Rod Home and John Heilbron who read the manuscript carefully and made the most useful comments and corrections. This book owes much to recent works on Nishina by Morris Low, Shizue Hinokawa, Kenji Ito and Dong Hoon Oh, with whom I had the pleasure to discuss our common interest. I thank Ito, Oh, Boumsoung Kim, and Hyungsub Choi who helped me to find some manuscripts and papers. The Department of the History of Science and Technology at Johns Hopkins University gave me two opportunities to read excerpts from the work in progress.

The staff of the Nishina Memorial Foundation, Institute of Physical and Chemical Research, Niels Bohr Archive, Special Collections of North Carolina State University, American Institute of Physics, Caltech Institute Archives, and the Bancroft Library, University of California, Berkeley, gave me access to many valuable manuscripts and photographs for the book, and I am grateful to them. I am especially grateful to Ms. Topsy Neher Smalley, the daughter of H. Victor Neher, who generously provided me with Neher's 1935 journal describing his trip to Japan and some photographs.

John Navas, first at IOP and then Taylor & Francis, has been my principal contact for the publication of the book, and I would like to extend my special thanks to him. I have been in poor health for some years and he has been most patient with my several delays of the final draft. The staff of Taylor & Francis has been also very helpful in preparing the book for publication.

Lastly but not least, I would like to thank my family for their warm encouragement.

Introduction

Japanese scientists and historians of science honor Yoshio Nishina as the "pioneer of modern physics in Japan," "founder of modern science in Japan," or as the "Christopher Columbus of Japanese physics." Hideki Yukawa, the first Japanese Nobel Laureate (physics, 1949), stated that without Nishina his generation could not have achieved such brilliant success in elementary particle physics during the 1930s and thereafter. Another Japanese Nobel Laureate (physics, 1965), Sin-itiro Tomonaga, made the same point in his eulogy of Nishina in 1951: "He made us aware of the modern methods of physical research." Eri Yagi concluded his article on Nishina for the *Dictionary of Scientific Biography* as follows: "Without Nishina's return from Europe with the principles of quantum mechanics, these two physicists [Yukawa and Tomonaga] might never have developed their potentials to the fullest."[1]

In 1991 there were several celebrations for the centenary of Nishina's birth: an international symposium was held in Tokyo to discuss Nishina's contributions to twentieth-century physics; a special documentary videotape with the title, "Nishina Yoshio: Father of Modern Physics," was coproduced by his hometown and the Nishina Memorial Foundation; and a special stamp was issued by the Japanese postal service on his birthday (December 6). Similarly, in 2005, Toshimitsu Yamazaki (Tokyo University and Riken) paid a tribute to the "Father of Nuclear and Particle Physics in Japan" with his paper, "Yoshio Nishina and the Dawn of Nuclear and Particle Physics," in a special session for the 50th anniversary of the establishment of the Nishina Memorial Foundation.[2]

I first encountered Yoshio Nishina in 1992 while examining the Cavendish Laboratory's annual photographs for my book on the history of the Cavendish. As I looked at the image of the young Nishina in the 1922 annual photograph, I vaguely remembered his name and asked myself, "What was this Japanese doing in this center of experimental physics?" I soon learned from fragmentary English sources that Nishina was trained under Niels Bohr for several years, that he coauthored the famous Klein–Nishina formula, and that during the 1930s and 1940s he contributed significantly to the development of several branches of physics in Japan. Later, after examining Japanese sources commenting on Nishina, I came to believe that I should investigate Nishina's life and works further. Thus I began what I considered at the time to be a "little" research project.

I quickly found that appraising Nishina's role in the international physics community and in the Japanese scientific community was more difficult than I had anticipated. The English sources hinted that the Klein–Nishina formula might be Nishina's only distinguished contribution to physics. For example, the *Dictionary of Scientific Biography* did not have an article for Nishina in its first edition in 1973.

It only added a short article for him in its supplementary volume in 1990 even though Nishina had died in 1951.[3] Therefore I struggled for some time to justify why I should study this seemingly "less important" physicist. On the other hand, as the first paragraph of the introduction indicates, the Japanese sources unanimously hailed him as a great scientist and as a perfect teacher. If so, why has the international scientific community and Western historians of science neglected him for so long? Getting to the bottom of this puzzle became one of the reasons for writing this biography of Yoshio Nishina.

I started the project to answer the following three questions: what kind of scientist (or physicist) was Nishina?; how good a physicist was he?; and how much and in what way did he contribute to the development of twentieth-century physics in Japan and the world? To answer these questions, I concentrated on analyzing the scientific works of Nishina and his junior researchers in the Institute of Physical and Chemical Research (Riken). I will argue that Nishina assumed three roles: a very competent researcher, a formidable teacher, and a shrewd administrator. By performing these three different, but closely related, roles magnificently, he not only made a significant contribution to the emergence and growth of a research network that eventually produced two Nobel Prize winners, but also raised the level of Japanese physics overall.[4]

While writing the biography, I have met several difficulties, most of which resulted from the simple fact that Nishina was a Japanese physicist. Clearly Nishina was Japanese. Thus, the social and cultural environments in which he had worked and had trained the subsequent generations of Japanese physicists were quite different from those in the West. Although Japan was a successful example of industrialization and Westernization in the late nineteenth and early twentieth centuries, it was still an East Asian country in which very different traditions dictated people's everyday life and thought. The teacher–student relationship and the funding system for research, for example, were not what they were in the West. In Japan, the prefix of "professor" meant not only a professional title but also the embodiment of respect and authority: therefore Yukawa and Tomonaga often called Nishina as "Professor Nishina" in their memoirs, although Nishina had never been appointed professor at any Japanese university. The question to be answered is therefore: how much did being Japanese influence Nishina's scientific work and contribute to his success?

On the other hand, Nishina was a physicist who had been deeply influenced by the Western research environment. After graduating from the Westernized Tokyo Imperial University and doing some research at Riken, he was thoroughly trained in Europe for more than 7 years at two leading research institutes: the Cavendish Laboratory in Cambridge and Niels Bohr's Institute of Theoretical Physics in Copenhagen. His most famous work, the Klein–Nishina formula, was produced while he was still working in Europe. Nishina befriended many distinguished physicists in Europe, including several Nobel Prize winners, who treated him as their equal. He spoke German and Danish fluently and English fairly well. Also Nishina was widely perceived as introducing the new quantum mechanics and a new research environment into Japan. Thus Nishina was one of the most Westernized Japanese physicists in the first half of the twentieth century. How, then, could he successfully manage his Western traits in the Japanese environment?

Perhaps a partial explanation is Nishina's unique quality that he described as "a fairly elastic adaptability to new conditions."[5] On the one hand, he was typically Japanese. He had been a model student from elementary school to university. He had behaved exactly as the typical Japanese did when he met and dealt with senior Japanese scientists or nonscientists outside his laboratory. That was why during the 1930s he excelled as the secretary of the 10th subcommittee of the Japan Society for the Promotion of Scientific Research that subsidized his cosmic ray research and later the construction of two cyclotrons; why he became the spokesman of science during World War II; and why he was regarded as the statesman of Japanese science during the American occupation that followed. On the other hand, Nishina behaved just like the leading Western scientists in his laboratory at Riken: he produced quality work in new fields of physics, became part of the international network, acted in an unauthoritarian manner, partook in open discussion and collaboration with junior researchers, and organized research groups of junior researchers with different educational backgrounds. Owing to this symbiosis of his Japanese and Western attitudes, Nishina enjoyed unanimous and unprecedented respect from both physicists and nonscientists.

Nishina's being a Japanese physicist also raises some problems of interpretation. In East Asia, such celebrated figures as Nishina traditionally are depicted as "perfect" men, flawless leaders whose mistakes or failures (if any) were not due to faults of their own. Criticism of heroes, particularly great *sensei* [teachers], is not permitted. Most Japanese sources on Nishina faithfully follow this tradition. Under such circumstances, one could doubt whether a historian of science who is culturally East Asian (as I am) can achieve an objective appraisal of Nishina's contributions. I agree with my Japanese colleagues that Nishina was a great man. Yet my appreciation of his greatness is based on ideas and opinions that are not necessarily those expressed in traditional Japanese scholarship, particularly in areas on how and why Nishina's role and contributions were critical to the development of the Japanese physics community. In studying Nishina's life and work, my goal has been to appraise his contributions accurately, and any such evaluation, honestly attempted, runs the risk of being somewhat critical.

Another problem of interpretation is that many Western physicists and historians of science regard the success of Japanese physicists since the 1930s as the exception rather than the rule. Perhaps as a result, Westerners have made little effort to incorporate the accomplishments of Japanese physicists into the larger framework of the history of physics. Worse, when studying the history of science in Japan, Westerners have tended to concentrate on a few topics such as Nishina's cooperation with Klein in the calculation of the Compton scattering cross section. They, however, often neglected Nishina's role in the spread of the fledging field of quantum mechanics in Japan and the construction of two cyclotrons. It should be emphasized that the latter was far more than a simple transfer of know-how from Ernest O. Lawrence's Radiation Laboratory in Berkeley to Riken.

Although Western scholarship has begun to recognize the merits of some twentieth-century Japanese physicists, in particular theoretical physicists like Yukawa and Tomonaga, the Japanese physics community in which those scientists worked has received scant attention.[6] Experimental works by Japanese physicists in the first half of the twentieth century has never been explored seriously by any Western scholars. Nor have Westerners endeavored to understand the Japanese physics community of the twentieth century within its broader intellectual context. Most fail to appreciate the community's independence and importance, too. Language barriers undoubtedly were an obstacle but in fact they are a convenient excuse since many important scientific works by Japanese physicists were published in English or other Western languages in either prestigious Western or Japanese journals.

This short biography of Yoshio Nishina certainly does not solve all these problems mentioned above. However, if the book provides readers with new perspectives and more questions, I will have achieved my goal. I sincerely hope that more extensive biographies of Nishina with fresh perspectives will appear in the near future. Yoshio Nishina certainly has the greatness to deserve not just one but many biographies.

NOTES

1 Masao Suzuki and Ryogo Kubo (eds.), *Evolutionary Trends in the Physical Sciences: Proceedings of the Yoshio Nishina Centennial Symposium, Tokyo, Japan, December 5–7, 1990* (Berlin: Springer-Verlag, 1991); Sin-itiro Tomonaga, "Dr. Nishina," in Makinosuke Matsui and Hiroshi Ezawa (eds.), *Sin-itiro Tomonaga: Life of a Japanese Physicist*, translated by Cheryl Fujimoto and Takako Sano (Tokyo: MYU, 1995), pp. 11–114 on p. 114; Hideki Yukawa interview with John A. Wheeler (July 10, 1962), AIP MSS OH 575, Eri Yagi, "Nishina, Yoshio," *Dictionary of Scientific Biography, Supplementary II* (New York: Charles Scribner's Sons, 1990), Vol. 18, pp.684–687 on p. 686.

2 Satosho, Okayama Prefecture and Nishina Memorial Foundation, *Nishina Yoshio: Father of Modern Physics (VHS)* (Okayama: Sanyo-eiga, 1991); Toshimitsu Yamazaki, "Yoshio Nishina and the Dawn of Nuclear and Particle Physics," in 2005 2nd Joint Meeting of the Nuclear Physics Divisions of the American Physical Society and the Physical Society of Japan (Session 3S: Nishina Commemorative).

3 *Dictionary of Scientific Biography* has a rule that only deceased scientists with distinguish achievements be considered to be included in it.

4 Dong-Won Kim, "The Emergence of Theoretical Physics in Japan: Japanese Physics Community between the Two World Wars," *Annals of Science, 52* (1995), 383–402.

5 Y. Nishina to N. Bohr (April 1, 1929) in *Supplement to the Publications No. 17, No. 20 and No. 21* (Tokyo: Nishina Memorial Foundation, 1986), pp. 6–8 on p. 7.

6 Laurie M. Brown and Olivier Darrigol have published a series of papers and books on Japanese physicists' work on elementary particle physics. For more information see Chapter 4.

List of Abbreviations

AIP	American Institute of Physics
EOL	Ernest Orlando Lawrence Papers, The Bancroft Library, University of California, Berkeley
Nishina MSS	Yoshio Nishina Manuscript, Institute of Physical and Chemical Research
Riken (or RIKEN)	Institute of Physical and Chemical Research
Special Collections, NCSU Libraries	Harry Charles Kelly Papers, 1882–1995, MC 72, Special Collections Research Center, North Carolina State University Libraries
Bulletin IPCR	*Bulletin of the Institute of Physical and Chemical Research*
DSB	*Dictionary of Scientific Biography* (New York: Charles Scribner's Sons, 1973)
Proc. Phys. Math. Soc.	*Proceedings of the Physico-Mathematical Society of Japan*
SP	*Scientific Papers of the Institute of Physical and Chemical Research*

Contents

1 Youth

1.1 THE NISHINA FAMILY

Yoshio Nishina was born on December 6, 1890 in Hamanaka, a hamlet of the village of Shinjo, today known as Satosho Village, of Okayama Prefecture.[1] The Okayama Prefecture is located between Kobe and Hiroshima along the Setonaikai sea in the western region of Japan's main island. His family had gained local distinction in the 1830s, when Yoshio's grandfather, Arimoto, used his civil engineering skills to settle a dispute between Shinjo and a neighboring village over possession of some salt-making fields. In recognition of this service, the local daimyo (feudal lord) bestowed samurai status on Arimoto and one of his sons, Arihito. Yoshio's father, Arimasa, was the fourth son of Arimoto and inherited some of the family's farming and salt-making fields. He married Tsune, a daughter of the headman of a distant town, Takafuta, in Hiroshima Prefecture. The couple had five sons and four daughters (Yoshio being the eighth). Arimasa died when Yoshio was 16 years old. Yoshio's eldest brother, Teisaku, 20 years older than Yoshio, inherited the house, salt-making and farming business, and responsibilities as head of the family, including the role of father to his siblings. Yoshio's second eldest brother, Empei, was an inventor. Yasuo, the third son in line, was an electrical engineer. Masamichi, 3 years younger than Yoshio, died young in 1919. Three of Yoshio's four sisters married their cousins. The youngest, Toku, who was very close to Yoshio, married into a successful industrial family, the Uchida family of the nearby city of Kurashiki. The Uchida family later would financially support Nishina's stay in Europe.

Yoshio was born in the heart of one of the most glorious and turbulent times in Japanese history, the Meiji period (1868–1912). This was the period during which Japan transformed itself into a modern, industrialized nation.[2] A strong new central government in Tokyo abolished the old feudal system and enthusiastically adopted Western-style systems of politics, economics, education, science, and technology. Japan's new cabinet and parliament ran relatively smoothly. Its infant educational and public health systems quickly surpassed those of many Western countries, and its new railway and telegraph lines rapidly networked the country. Japanese industry began to dominate trade in the Far East and Southeast Asia by manufacturing and exporting products that, although less technologically sophisticated than Western products, were more familiar and thus more acceptable to Asian consumers. Japan's most surprising development in the eyes of the rest of the world was its emergence as a major military power in Asia. In 1894–1895 Japan's modernized armed forces defeated the Chinese army and navy on the Korean peninsula and then, in 1904–1905, vanquished Russian forces on land in Manchuria and at sea in the Tsushima Straight between Korea and Japan.

In this rapidly changing environment, a family's fortune rose or fell depending on its members' ability to adapt. The Nishina family's fortune was hamstrung by its ties to the dying feudal system. Thus, the local paper currency that the Nishina family had issued for local use became worthless when the central government issued the new national currency. The failure of the family-owned company accelerated the decline of Nishina family's fortune.[3] In December 1907, Empei received a patent for his fire-proof paint, which earned praise from the influential *Yomiuri Shimbun* (*Yomiuri Newspaper*) in the following year. Teisaku, head of the family, decided to set up a company, Nishina & Co., to manufacture and sell Empei's paint and other inventions, and invested a good part of the family fortune in the company. Empei's products, however, proved ineffective and the business rapidly declined. By late 1913, money had become so tight for the Nishina family that Teisaku was unable to send Yoshio his monthly allowance regularly.[4]

To restore their fortune, the Nishina family looked to the younger generation, and particularly to Yoshio, the family's most brilliant member. This expectation, as asserted by the Japanese historian of science, Kenji Ito, greatly influenced Nishina's decisions, especially his choice of career.

> Nishina Yoshio's great goal was to restore the house of Nishina. His search for success, his training to be an engineer, his entrepreneurial ambitions — all aimed at a restoration of the family fortune of the Nishina clan.[5]

However, Yoshio did not fulfil his family's hope that he would rebuild their fortune. Instead, his great contribution was made to the Japanese physics community.

1.2 YOSHIO NISHINA: THE STUDENT

When Yoshio entered school at the age of seven, he became a model student and remained so throughout his years of formal education.[6] As an elementary school student in his native village of Shinjo, he earned straight A's in subjects that included Japanese, arithmetic, and ethics. His performance at Shinjo's higher elementary school, which he entered in 1901, and then at the neighboring Kamogata's higher elementary school, to which he moved in 1904 after the Shinjo school was closed, earned Yoshio a graduation award for academic distinction from the prefectural governor in March 1905. His academic record for the next five years of schooling in Okayama, to which he moved to enter that city's middle school, was unmatched in the school's history: he earned A's in every subject every year, ironically, except in physics in his fifth year (Figure 1.1).

A letter that Yoshio wrote to his younger brother Masamichi in 1910 sheds light on Yoshio's attitude toward hard work.[7] In that letter, Yoshio advised his brother to prepare for every class on the previous day, to listen carefully in class, to review the material on the following day, and to review all subjects during the weekend, just as he himself had done in the middle school. Yoshio also found time for sports in middle school: tennis became his favorite and life-long hobby. Yoshio's stellar performance at Okayama Middle School, from which he graduated in March 1910, qualified him for acceptance by more prestigious high schools in Tokyo and Kyoto,

FIGURE 1.1 Yoshio Nishina in Youth. Above: Nishina's birth place in Shinjo Village (now Satosho Village) of Okayama Prefecture (Photo taken by the author). Below left: Yoshio (left in the back row), Yasuo (an elder brother, middle in the back row), Masamichi (younger brother, right in the back row) and children from the Nishina family. Below right: Yoshio as fifth year middle school student. (Courtesy of the Special Collections, NCSU Libraries.)

but Yoshio elected to enroll in the high school in Okayama because of its proximity to his hometown.

The Sixth High School in Okayama admitted Yoshio in April 1910 without requiring him to take the usual entrance examination. The high school was noted for its extraordinary discipline. Its principal, a retired army captain named Sentaro Kaneko, maintained a Spartan environment in which "samurai-like simplicity and fortitude were valued and loyalty to the state and emperor was unquestioned."[8] Nishina continued being the top in his class during his years at the Sixth High School.

Yoshio's entrance to high school presented him with the important decision of selecting his major subject of study. Among the options available, he excluded law, literature, science, and agriculture because they did not guarantee financial success in the Japanese society of the day. He considered medicine, but then abandoned the idea because a botched operation in his childhood had deprived him of his sense of smell. Engineering thus seemed the best choice. However, before his second year began in September 1911, Yoshio had to request a year's leave from school because of the pleurisy.[9]

As his high school graduation approached, Yoshio again had to choose his major field of study, this time for his university undergraduate. Yoshio's personal tastes clearly inclined toward scholarship but it was hardly the path most likely to restore his family's fortune. He approached his elder brothers for advice. Teisaku analyzed Yoshio's four possible career choices: scholar, civil servant, businessman, and engineer.[10] Although Yoshio had the qualities of a good scholar, Teisaku warned him that scholarship would not bring him sufficient income. Since civil service and business were not good matches with Yoshio's talents, this left engineering, a field Teisaku described as much closer to scholarship, "something between" civil service and business. Empei, the inventor, advised Yoshio that electrical, civil, and mining engineering were well-paying professions but that, if Yoshio wanted to become a scholar, mechanical engineering would be the best choice.[11] Yasuo, the electrical engineer, considered civil and mining engineering as the most lucrative career choices and mechanical and electrical engineering as the most intellectually stimulating.[12] Yoshio applied to the Department of Electrical Engineering at Tokyo Imperial University, and was accepted. However, Yasuo, who had reconsidered his advice, recommended civil engineering as the best choice, with electrical engineering as only the next best. He warned Yoshio:

> In any case, the goal is to gain fame and fortune .... So, you only need to choose the best career to promote yourself.... If you wish to become a scholar, electrical engineering would be the most interesting, since it is a new field with a lot of possibility to make new inventions. However, a scholar must be prepared to give up money.[13]

Concerned that his decision to study electrical engineering would hamper his prospects for restoring the family fortune, Yoshio immediately sent a plea to the dean of the College of Engineering asking for a transfer from the Department of Electrical Engineering to the Department of Civil Engineering because of "family matters."[14] The dean however denied his request.

In September 1914 Yoshio entered Tokyo Imperial University. A flare-up of his pleurisy in the spring of 1915 made it impossible for him to take his final examination in June, so he had to repeat his first academic year. Again he excelled in his studies during his first academic year at the university (1915–1916) ranking second among

43 students in the Department of Electrical Engineering with a grade point average of 92.4.[15] In the next academic year, he was the first.

Tokyo Imperial University's Department of Electrical Engineering provided Nishina with a thorough grounding in theory and also rigorous practical training. Ito has summarized the three-year program as follows:

> The first year focused on fundamental skills and knowledge necessary to electrical engineering. A little less than half of the class time was devoted to lectures. The principal subjects included electromagnetism, mathematics, dynamics, and thermodynamics. In the second year, more practical subjects came into the curriculum, and lecture courses related to power network occupied the largest portion of the instruction. Such courses included design of electric plants, power transmission and distribution, electrical lighting, design of generators, dynamos, and converters. Although not given as much time, alternating current theory and electrical railroad were taught in the second year. There was little formal instruction in the third year, which students would typically use to write their thesis.
>
> The lecture-style instruction took less than a half of the class time. In the first year, drawing and experiment occupied 23% and 30% [respectively] of the total class time of forty hours per week on average. Drawing was obviously crucially important for future engineers. They learned not only projective geometry but also how to draft actual designs of electrical engineering products. The Department took the training in drawing very seriously. Students were required to have their works book-bound and submitted to the Department, of which library would preserve their *études* of drawings permanently along with their thesis.
>
> . . .
>
> In the second and third years, the students were required to take a "practical training." The experience of practical training differed among students, it specifically being decided in consultation with the academic adviser. . . . For his third year, [Nishina] commuted to the Testing Department of Shibaura Engineering Works from January 15, 1919, working everyday from 7 to 5.[16]

Nishina's academic adviser was the energetic Hiderato Ho, who, despite being the youngest full professors in the department, was the most influential. Ho's emphasis on "graphical or pictorial understanding of electrical phenomena" and his "theoretical approach to electrical engineering" deeply influenced generations of students.[17] Nishina credited Ho with teaching him "how to grasp the physical meaning of things."[18] Ho authored a book on the theory of alternating current, one of the four books that most influenced Nishina during his undergraduate studies. The other three books were James H. Jeans' *The Mathematical Theory of Electricity and Magnetism*, Charles P. Steinmetz' *Theory of Calculation of Alternating Current Phenomena*, and Engelbert Arnold's *Wechselstromtechnik*.[19] It was Ho who suggested Nishina's thesis topic, a theoretical investigation of the three-phase alternating current generator, motor, and transformer, and who arranged for Nishina to work and collect data for his thesis in the testing laboratory of Shibaura Engineering Works.[20]

Impressed by Nishina's ability, Shibaura Engineering offered Nishina a permanent position even before he graduated, but Nishina envisioned a different career path for himself and declined the offer. A few months before graduation, he wrote a letter to his eldest brother Teisaku: "Since electrical engineering already has been fully developed, I would like to study electro-chemistry. I am planning to study chemistry in graduate school for a year and then . . . start practical work."[21] When Nishina graduated in the summer of 1918, he was among the handful of top students who were awarded silver watches by the Emperor. He also was elected to represent the College of Engineering graduates in the graduation ceremony, an honor that he was forced to decline because he was unable to attend the rehearsal.[22]

In the fall of 1918, Nishina entered Tokyo Imperial University's Graduate School of Engineering. His refusal of Shibaura Engineering's job offer had soured his relationship with Ho.[23] He therefore turned for guidance to another young electrical engineering professor, Tsunetaro Kujirai, an insulating materials specialist who had just returned from study in Europe. At the same time, Nishina entered his new adviser's laboratory in the Institute of Physical and Chemical Research (Riken).

Immediately after entering graduate school, Nishina's interest turned from electrochemistry to physics. His notebook began to be filled with Einstein's theory of relativity along with diagrams of triode vacuum tubes.[24] He thought that "most problems in dynamo[s] and motor[s] were already solved," but physics intriguingly "seemed full of unsolved puzzles."[25] A conversation with Hantaro Nagaoka in the fall of 1918 prompted Nishina to attend Nagaoka's lectures and to try some physics experiments in Nagaoka's laboratory. Captured by these experiences, Nishina immersed himself in physics under Nagaoka's tutelage until the spring of 1921.[26]

In gaining Nagaoka as his patron and Riken as the site of his research, Nishina was doubly fortunate. Nagaoka, famous for his Saturnian-ring atomic model, was the most prominent Japanese physicist of the early twentieth century. Riken, Japan's pre-eminent research institute during the interwar period, would send Nishina to Europe to study and later offer him his own laboratory, which Nishina would develop into a renowned center for theoretical, cosmic-ray, and nuclear research.

1.3 INSTITUTE OF PHYSICAL AND CHEMICAL RESEARCH (RIKEN)

The establishment of Riken was the combined result of the growth of Japanese science and technology and Japan's rapidly changing milieu during the mid-1910s. Ever since Japan had opened its doors to the rest of the world in 1853, Japanese scientists and engineers had imitated Western science and technology, but neither government nor industry had made a systematic effort to support scientific research. The desire for basic research emerged in academic circles at the beginning of the twentieth century but the time was not yet ripe. In 1913, Jokichi Takamine, an eminent applied chemist, argued persuasively that Japanese industry must be based on science, and proposed that Japan establish a "national science institute" funded by an initial investment of 20 million yen, a sum equal to the cost of one new battleship.[27]

Takamine's pioneering proposal at first fell mostly on deaf ears in government and industry circles, but the outbreak of World War I in Europe in 1914 dramatically changed Japan's attitude toward the importance of scientific research.

The economic blockade of Germany imposed by Britain and France hurt Japanese society immediately and badly, underscoring the dangers of being dependent on foreign technologies and research. Historian of science James R. Bartholomew outlined the war's contribution to the creation of Riken:

> World War I posed a major challenge to Japanese science, as it did to science worldwide.... In Japan the war heightened interest in the physical sciences. Indeed, one chemist called it a "blessing from heaven." Nonacademic research flourished there as never before. A significant shift of research activity into private firms and institutions took place. Both public and private sectors showed more inclination to spend money on science at home. Businessmen and officials who had earlier disdained research now became its promoters. Costly projects like the Institute of Physical and Chemical Research, which could not have gained a foothold before 1914, now moved to the top of the country's agenda.
>
> The blockade of Germany by Britain and France helped to produce these results. Before the summer of 1914, Japan had depended heavily on Germany for industrial chemicals, pharmaceuticals and precision instruments. All of the Salvarsan 606 used to treat syphilis came from Germany, and most of the thirty-four million yen spent on imported pharmaceuticals also went to that country. Much of the information used in producing aniline dyes was obtained from German-held patents. About 150 Japanese traveled to Germany each year for medical studies. Two years of every three spent abroad by Japanese scientists were spent in Germany. And Germany was the site favored by most Japanese for exhibitions and academic conferences. The blockade created a crisis. Prices of German-made products soared—when one could obtain the products at all. In a matter of weeks, aniline dyes jumped to twenty times their previous price. Serious shortages developed. By late September Tokyo reportedly had a six months' supply of most basic medicines, but Osaka had nearly run out.[28]

To meet these grave challenges, Joji Sakurai, an eminent physical chemist, and Eiichi Shibusawa, an influential financier, again brought up the need to create a national institute that would devote itself to basic research.[29] Physicists soon joined chemists in promoting the idea of a central research institute, and together the two groups secured support from industry and from powerful national politicians, including the Prime Minister Shigenobu Okuma. A special government committee was appointed to draft an official proposal, which was submitted to the Diet in the spring of 1915. This document emphasized that major Western countries already had established national laboratories for basic science research: Great Britain had its National Physical Laboratory, the United States its National Bureau of Standards, and Germany its Physikalische Technische Reichsanstalt as well as Kaiser Wilhelm Gesellschaft.[30] The proposal's final draft of 1916 clearly stated the rationale and goals of the proposed institute:

> Wishing to contribute to world civilization, to enhance the status of our nation, to lay the foundations of various industries and to increase the nation's wealth, it is imperative that we encourage creative research in the disciplines of physics and chemistry. The recent war, moreover, has taught us the urgent necessity of independence and self-sufficiency in military supplies and industrial materials and has made us all acutely aware of the need for physical and chemical research. Because these kinds of research facilities have not

> previously existed in our country, certain public spirited people are hoping to establish the Institute of Physical and Chemical Research.[31]

Once the proposal was finalized, progress toward establishment of Riken rapidly accelerated. In March 1916, the Diet passed the final draft of the proposal, along with the proposed subsidies amounting to 2 million yen over a period of 10 years. Japanese industry pledged to donate over 2 million yen, which however was disappointingly far short of the original goal of 5 million yen. In March 1917, a formal application to establish Riken was submitted to the government and was approved in three days. Next, Riken's first director, the eminent mathematician and educator, Dairoku Kikuchi, was appointed. In April 1917, the imperial family donated 1 million yen, adding great prestige and momentum to the effort to develop this so-called "private" institute.[32] In September 1917, as construction began at Riken's proposed site in Tokyo's Komagome area, basic research projects commenced in several university laboratories.

A detailed list of Riken's projected activities, published in January 1917, underscored the wide scope of Riken's objectives in comparison with those of Japan's other science and engineering laboratories.[33] These ambitious objectives included conducting pure scientific research and connecting it with applied research, carrying out research requested by government agencies, linking Japan's research institutes and testing laboratories, training researchers, providing outside researchers with free use of Riken's facilities, subsidizing research, encouraging inventions, publishing research results, and sponsoring public meetings.

Carrying out these important missions obviously required a continuous flow of enormous funding, but, after the end of World War I in the fall of 1918, obtaining such funding proved difficult.[34] In the competitive post-war Japanese market, Western products that were less expensive than Japanese goods and also of higher quality flooded again. Japanese companies resumed business with Western companies to acquire the advanced technology and goods that they needed. The wartime goal of self-sufficiency that had propelled the establishment of Riken was almost forgotten, and Japanese industry's formerly enthusiastic support for this goal cooled. Riken's operating budget was slashed to "between 120,000 and 130,000 yen . . . one-third the sum provided in the 1917 plan, and only about one-eighth the amount originally suggested by Takamine."[35] Moreover, rising material and labor costs tripled the estimated costs of constructing the Riken facilities. In the face of these financial difficulties, Riken's researchers labored to continue their research and achieved some notable successes. In 1919, for example, Umetaro Suzuki began research on "compound sake," a nontraditional type of sake made from materials other than rice. By 1921, research was being conducted for its production by the newly established Yamato Experimental Distillery. In 1922, Katsumi Takahashi isolated vitamin A from cod-liver oil, which Japanese subsequently called "Riken Vitamin."[36]

On September 30, 1921, Masatoshi Okochi became Riken's third director, succeeding Koi Furuichi (1917–1921).[37] A member of the aristocracy, Okochi had assumed the title of baron in 1907. He had majored in armament engineering at Tokyo Imperial University, where in 1911 he was appointed professor. When Riken was established in 1917, he and Nagaoka shared responsibility for directing the physics section.

FIGURE 1.2 The Building No. 3 of the Institute of Physical and Chemical Research (Riken) where Nishina would open his own laboratory in 1931. Riken was the most important research institute during the interwar period in Japan, largely owing to its third director, Masatosh Okochi (right). (Courtesy of the Institute of Physical and Chemical Research.)

With strong allies in the Diet and government, Okochi was the ideal choice to save the troubled infant institute. One of his first actions as Riken's director was to abolish the institute's antagonistic two-section structure consisting of a chemistry section and a physics section, which had been fighting each other for control of Riken since its inception. Okochi replaced this system with a structure of independent laboratories, each of which was directed by a head with full discretionary power. In the Japanese system in which the work of subordinates was strictly controlled, this institutional innovation greatly enhanced research freedom at Riken by enabling young laboratory heads to pursue their own research agendas without interference from their superiors (Figure 1.2).

Okochi's most important contribution to Riken's development, however, was to secure stable funding for the institute's research. His revolutionary solution to Riken's financial problems was, in the words of Tessa Morris-Suzuki, "for the institute to commercialize its own inventions, thus earning revenue to support future research activities." As Morris-Suzuki stated:

> In 1927 Okochi established the Physical and Chemistry Industrial Promotion Company (Rikagaku Kogyo, KK) with the research institute as major shareholder but with contributions from a number of other large investors, including the Mitsui, Mitsubishi and Sumimoto zaibatsu. The institute's corporate arm acted as a holding company, providing what would now be called "venture capital" to a mass of industrial firms created to apply the fruits of the Riken research projects. Within twelve years Rikagaku Kogyo had become the center of a web of over sixty enterprises, extending throughout Manchuria and the Korean peninsula as well as the Japanese archipelago itself. What had begun as a scientific research institute had grown into an industrial conglomerate.[38]

These venture companies used Riken's patents to manufacture alumite, positive paper, piston rings, and other products. Thanks to the steady stream of income from Rikagaku

TABLE 1.1
Short Summary of Finance of Riken

Year	Government subsidies	Income from KK	Total income (A)	Research expense	Total expense (B)	(A) − (B)
1921	—	—	306	203	249	57
1922	—	—	319	290	320	−1
1923	250	—	943	754	900	43
1924	250	—	886	493	886	0
1925	250	—	987	453	988	−1
1926	250	—	1181	471	1181	0
1927	250	—	660	568	660	0
1928	250	n/a	855	624	669	186
1929	250	n/a	817	677	721	96
1930	250	n/a	653	609	653	0
1931	250	n/a	652	604	647	5
1932	250	n/a	684	637	684	0
1933	250	n/a	843	775	829	14
1934	250	162	840	744	823	17
1935	150	219	937	835	915	22
1936	150	1142	2150	1158	1240	910
1937	150	1531	2387	1439	1537	850
1938	—	2599	3074	1786	1889	1185
1939	—	3033	3705	2311	2420	1285
1940	—	n/a	3611	2901	3071	540
1941	—	n/a	3598	2939	3063	535

Unit: 1000 yen; n/a: not available.

Source: Satoshi Saito, "The Finance of the Institute of Physical and Chemical Research and the Research Funds," *Butsuri*, *45* (1990), 761–765 on 761.

Kogyo, Riken's financial condition improved dramatically from the mid-1930s on, as shown in Table 1.1.

With its financial stability assured, Riken could pursue one of its primary missions: to conduct pure scientific research. Nishina would become the major beneficiary of Riken's largesse, which would make possible his researches on quantum physics, cosmic rays, and nuclear physics. Notably, Nishina's construction of the 26-inch cyclotron began in 1935, just as Riken's financial situation took a turn for the better.

1.4 JAPAN'S PHYSICS COMMUNITY IN THE EARLY TWENTIETH CENTURY

In comparison with chemistry, biology, and medicine in the first decades of the twentieth century, physics was a small, neglected field in Japan.[39] Tokyo Imperial University's Department of Physics, which since 1876 had trained and produced most of Japan's physicists, had only two full professors at the end of the nineteenth century. In the nine years between 1898 and 1907, the University produced only 78 physicists.[40] In the five years between 1902 and 1907, Kyoto Imperial University produced a mere 16 physicists. Research had not yet become an activity expected of Japanese physicists, and during the period between 1898 and 1907 only seven Japanese physicists published more than 10 papers.[41] In the early 1900s, Japan's only serious physics researcher was Nagaoka, who surpassed other Japanese physicists both in the quality and quantity of his publications. The most famous of his publications, a series of papers on the structure of the atom that included his famous Saturnian-ring model, were published from 1903 to 1905.[42]

A serious problem was that theoretical physics had not yet been assigned a proper role in Japanese science. The situation in Europe and America may have been similar but Japanese one was more serious.[43] Because most Japanese believed that physicists should investigate electricity and magnetism, heat, metallurgy, and other applied subjects to benefit Japan's emerging industrial sector, physicists focused their attention mostly to these "useful" fields. A survey of physics papers between 1898 and 1907 proves this point clearly: 122 research studies in magnetism were published, 26 in light, 20 in electricity, 20 in heat, 15 in sound and vibration, and 14 in instrumentation.[44] Seismology, gravity, and other geophysical subjects were popular research subjects for many Japanese physicists. Theoretical studies were significantly less frequent: atomic structure and radiation were investigated in only 10 published studies and mathematical physics in 8.[45] Even Nagaoka and Aikitsu Tanakadate, the two most influential figures in the Department of Physics at Tokyo Imperial University, were busy working on experimental and applied subjects.

The work of Kotaro Honda, "founder of the science of metals in Japan," exemplified the dominance of experimental and applied physics over theoretical and basic physics.[46] For most Japanese, Honda was the "symbol of what physicists would do for the welfare of the country."[47] Honda published several important papers on metallurgy in German and English journals and was awarded the prestigious Bessemer Medal for his research on magnetic properties of iron and steel in 1922. One of the major accomplishments of his long, distinguished career was the establishment of the Research Institute for Iron, Steel, and other Metals in 1919, which served both military and civilian shipbuilding.

Another idiosyncrasy of the Japanese physics community that obstructed the growth of physics was that theoretical physicists distanced themselves from experimental physics. More often than not, Japanese physicists regarded theoretical physics as "applied mathematics" and treated it as such.[48] In 1901 at Tokyo Imperial University, where most future physics faculty of Japan's other imperial universities were trained, theoretical and experimental physics were officially separated.[49] The flagship journal of physics in Japan, established under the name *Proceedings of the Physico-Mathematical Society in Japan* in 1901, also illustrates the intimate partnership between mathematics and theoretical physics.

The close cooperation between experimental and theoretical physics that flourished in Cambridge, Berlin, Göttingen, and Munich at the turn of the century therefore did not exist in Japan. The feudal tradition in the Japanese university system, where a professor monopolized a field and controlled the work of his subordinates precluded both collaboration and free exchange of ideas among physicists. Only Nishina was capable of breaking these barriers in his laboratory in Riken and thus ushering in a new era for Japanese physics.

NOTES

1 For Nishina's early life, see Tetsuo Tsuji, "The Life of Yoshio Nishina," *Butsuri*, *45* (1990), 712–719 on 712–714; Kenji Ito, *Making Sense of Ryoshiron (Quantum Theory): The Introduction of Quantum Mechanics into Japan, 1920–1940* (Ph.D. dissertation, Harvard University, 2002), Chapters 4 and 6.
2 For the history of Meiji period, see Marius B. Jansen, *The Making of Modern Japan* (Cambridge, MA: Harvard University Press, 2000), Chapters 11–14.
3 For Empei Nishina and the failure of Nishina & Co., see Ito, *Making Sense of Ryoshiron*, pp. 321–328 and 334.
4 Ibid., pp. 335–336.
5 Ibid., p. 338.
6 Tsuji, "The Life of Yoshio Nishina," 712–713.
7 Yoshio Nishina to Masamichi Nishina (April 7, 1910), in *The Collected Letters of Dr. Yoshio Nishina* [in Japanese] (Satosho: Nishina Memorial Foundation, 1993), pp. 28–40 on pp. 29–31.
8 Ito, *Making Sense of Ryoshiron*, p. 314.
9 The Sixth High School to Yoshio Nishina (September 27, 1911), in *The Collected Letters of Dr. Y. Nishina*, p. 66.
10 Teisaku Nishina to Yoshio Nishina (February 20, 1913), ibid., pp. 85–90 on pp. 85–87.
11 Empei Nishina to Yoshio Nishina (May 14, 1914), ibid., pp. 95–98 on pp. 96–97.
12 Yasuo Nishina to Yoshio Nishina (October 2, 1913), ibid., p. 91.
13 Yasuo Nishina to Yoshio Nishina (May 31, 1914), ibid., pp. 105–112 on pp. 107–108.
14 Yoshio Nishina to Wataru Watanabe (June 12, 1914), ibid., p. 113.
15 Yoshio Nishina to Tsune and Teisaku Nishina (July 8, 1916), ibid., pp. 135–137 on p. 135.
16 Ito, *Making Sense of Ryoshiron*, pp. 179–183.
17 Ibid., p. 184.
18 Y. Nishina, "What I Have Read: Remembering the Overseas Study Period," in *Atomic Power and I* [in Japanese] (Tokyo: Gakufu Shoten, 1950), pp. 224–230 on p. 226.
19 Ibid.
20 Y. Nishina, *Effect of Unbalanced Single-Phase Loads on Poly-Phase Machinery and Phase Balancing* (undergraduate thesis, Tokyo Imperial University, 1918), preface.
21 Yoshio Nishina to Teisaku Nishina (April 8, 1918), in *The Collected Letters of Dr. Y. Nishina*, pp. 138–140 on pp. 138–139.
22 Yoshio Nishina to Masamichi Nishina (July 10, 1918), ibid., pp. 145–146.
23 Yuichiro Nishina, "Yoshio Nishina in Recollection [in Japanese]," *Butsuri*, *45* (1990), 724–726 on 724. Ho became angry at Nishina since Nishina made such an important decision without consultation with him.
24 Ibid.
25 Nishina, "What I Have Read," pp. 226–227.
26 Ibid.
27 Kiyonobu Itakura and Eri Yagi, "The Japanese Research System and the Establishment of the Institute of Physical and Chemical Research," in Shigeru Nakayama, David L. Swain,

and Eri Yagi (eds.), *Science and Society in Modern Japan: Selected Historical Sources* (Cambridge, MA: MIT Press, 1974), pp. 158–201 on pp. 170–171.

28 James R. Bartholomew, *The Formation of Science in Japan: Building a Research Tradition* (New Haven: Yale University Press, 1989), p. 199.

29 For more information about the establishment of Riken, see Itakura and Yagi, "The Japanese Research System," pp. 182–192 and Bartholomew, *The Formation of Science in Japan*, pp. 212–217.

30 "The Proposal for the Establishment of the Institute of Physical and Chemical Research (April, 1915)," in *The History of Physics in Japan* [in Japanese] (Tokyo: Tokai University Press, 1978), Vol. 2, pp. 271–272.

31 Itakura and Yagi, "The Japanese Research System," pp. 188–189.

32 The official history of Riken notes that it started as a "private research foundation." See "History: Private Research Foundation Period/Corporation Period" (www.riken.jp/engn/r-world/riken/history).

33 "The Activities of the Institute and the Industry," in *The History of Physics in Japan*, Vol. 2, pp. 273–274.

34 Itakura and Yagi, "The Japanese Research System," pp. 192–194.

35 Ibid., p. 194.

36 Roundtable Discussion, "The Creative Minds of Chemists Revealed: From Riken Vitamin to Compound Sake [in Japanese]," *Shizen* (December, 1978), 28–37.

37 For the life and work of Masatoshi Okochi, see Kinenkai Okochi (ed.), *Okochi Masatoshi: Life and His Works* [in Japanese] (Tokyo: Nikkankogyo Shimbunsha, 1954).

38 Tessa Morris-Suzuki, *The Technological Transformation of Japan: From the Seventeenth to the Twenty-First Century* (Cambridge: Cambridge University Press, 1994), p. 127.

39 For the first generation of Japanese physicists, see Kenkichiro Koizumi, "The Emergence of Japan's First Physicists: 1868–1900," *Historical Studies in the Physical Sciences, 6* (1975), 3–108.

40 *The History of Physics in Japan*, Vol. 1, p. 111.

41 Ibid., pp. 152–153.

42 For Nagaoka's atomic models, see Eri Yagi, "On Nagaka's Saturnian Model (1903)," *Japanese Studies in the History of Science, 3* (1964), 29–47; Eri Yagi, "The Development of Nagaoka's Saturnian Atomic Model I: Dispersion on Light (1905)," *Japanese Studies in the History of Science, 6* (1967), 19–25; and Eri Yagi, "The Development of Nagaoka's Saturnian Atomic Model II (1904–05): Nagaoka's Theory of Structure of Matter," *Japanese Studies in the History of Science, 11* (1972), 73–89.

43 For the development of theoretical physics in Europe and America, see Christa Jungnickel and Russell McCormmach, *Intellectual Mastery of Nature: Theoretical Physics from Ohm to Einstein*, 2 volumes (Chicago: University of Chicago Press, 1986); Silvan S. Schweber, "The Empiricist Temper Regnant: Theoretical Physics in the United States," *Historical Studies in the Physical Sciences, 17* (1986), 55–98.

44 *The History of Physics in Japan*, Vol. 1, pp. 152–153 and 182–222.

45 Ibid.

46 For Kotaro Honda's life and works, see Nobuo Kawamiya, "Kotaro Honda: Founder of the Science of Metals in Japan," *Japanese Studies in the History of Sciences, 15* (1976), 145–158. For his laboratory in Tohoku Imperial University, see Bartholomew, *The Formation of Science in Japan*, pp. 187–191.

47 D. Kim, "The Emergence of Theoretical Physics in Japan," 386.

48 Ito, *Making Sense of Ryoshiron*, Chapter 2, Section 3 "Theoretical Physics in 'Theory'."

49 The two departments of physics merged again in 1920.

2 Nishina in Europe

> New physics was just being born in Europe. Europe was boiling with [the] expectation of new ideas and new discoveries. This lucky event matched [Nishina's] talent and ambition. Fate made him stay in Europe for seven years, which he himself probably did not originally plan. This transformed him into a first rate physicist, which would not have been really possible if he had remained in Japan.
>
> **Ryogo Kubo**[1]

> Indeed, in the laboratory we miss your experience and helpfulness constantly, and besides we are in our theoretical discussions almost daily reminded of the beautiful work you made just before you left in collaboration with Klein .... The famous formula which was the result of this work is not only, as I need not say, the basis for the interpretation of the scattering measurements in which we — not least due to Jacobsen's work — are intensely interested, but the striking confirmation which this formula has obtained became soon the main support for the essential correctness of Dirac's theory when it was apparently confronted with so many grave difficulties.
>
> **Niels Bohr to Yoshio Nishina**[2]

> It is just a fortnight ago since I left Copenhagen .... Owing to my inability I have not accomplished anything valuable in physics during this long stay.
>
> **Yoshio Nishina to Niels Bohr**[3]

2.1 HEADING FOR EUROPE

On April 5, 1921, Nishina boarded the *Kitanomaru* at Kobe and sailed to Europe. His voyage would last a month. Years later he told his younger son, Kojiro: "Many members of my family and relatives came to say good-bye. I could see your grandmother waving her hands until almost everyone left the dock. I can remember the scene vividly even now." Nishina was his mother's favorite, and she was so depressed by his departure that, upon her return home from the dock, she fell ill. She died 18 months later, on October 3, 1922, never seeing her beloved Yoshio again.[4]

Riken supported Nishina's overseas study through a special program to send its best researchers to European or American laboratories for advanced training. An article in *Nature* (1918) said: "A few of these associate fellows [junior researchers] are annually sent abroad for further training, there being three (Asahara, Nishikawa, and Takamine) in the United States at present."[5] The usual duration of study abroad was two years, during which the junior researchers were paid a special "overseas study" salary. If a researcher remained abroad longer than two years, this special salary was replaced by a basic salary. In either case, the amount was insufficient, and Riken generally expected researchers' families to contribute supplemental support for their upkeep. For the first three years of Nishina's study abroad, he received additional support from the family of his third sister's (Toku's) husband.[6]

During his study abroad, Nishina intended to continue research on x-ray spectroscopy and related subjects begun at Riken with Nagaoka. Nagaoka had shifted his research concentration to spectroscopy and radioactivity around the end of World War I, and had jointly published several papers with Yoshikatsu Sugiura on the spectra of mercury, nitrogen, carbon monoxide, bismuth, and radioactive isotopes.[7] In 1919, Nagaoka reported on the progress of this research in a letter to Ernest Rutherford, who in that year was named director of the Cavendish Laboratory of Cambridge University:

> As to the radioactive work in Japan, there was as yet none worth mentioning. The government has purchased for the use [of] my laboratory radium salt containing 100 mgrm of the metal, so that there is now [the] prospect of starting radioactive experiments. We are now going to investigate the spectrum of the emanation, examine the structure of the lines, and see if they are arranged in series. We are encountering great difficulty in the purification of the emanation.[8]

Nagaoka wished to accelerate this research by sending his best pupils, including Taiji Kikuchi, Nishina, and Sugiura, to Europe. His natural choice of destination for these young researchers was the world's foremost center for experimental physics, the Cavendish Laboratory. Taiji Kikuchi, Nagaoka's most promising protégé and a son of the Baron Dairoku Kikuchi, was the first whom Nagaoka chose to send to the Cavendish. Unfortunately, Kikuchi died prematurely in Cambridge in March of 1921, probably because of radioactive sickness. Under Nagaoka's influence, Riken chose Nishina to replace Kikuchi.

In May of 1921, Nishina finally arrived in Europe. After spending a few months on intensive English training, he matriculated as a member of Emmanuel College, and then, in October, entered the Cavendish Laboratory.

2.2 CAMBRIDGE AND GÖTTINGEN

By 1921, the Cavendish Laboratory had already produced six Nobel Prize winners in physics and chemistry. These were the Cavendish's second director, Lord Rayleigh (physics, 1904); its third director, Joseph John Thomson (physics, 1906); its fourth director, Ernest Rutherford (chemistry, 1908); the father–son team of William Henry and William Lawrence Braggs (physics, 1915); and Charles Glover Barkla (physics, 1918). Rutherford was a truly outstanding experimental physicist at the peak of a long, distinguished career. In 1898, while a research student working under Thomson, Rutherford was the first to distinguish α- and β-rays. Later, at McGill University in Montreal, Canada, he worked on the radioactivity of elements, which brought him the 1908 Nobel Prize in chemistry. Shortly before being awarded that honor, in 1907, he returned to England to accept a chair at the University of Manchester, where, in 1911, he ushered in the nuclear age with his discovery of the atom's nucleus. During World War I, he and Henry Bragg became the most important figures in Britain's war-related research. At the war's end, Rutherford immediately launched a new line of investigation into the artificial disintegration of the nucleus. Rutherford was also an exceptional teacher who attracted many brilliant young physicists. While he was in Montreal, he trained Otto Hahn; in Manchester, his pupils included, among many others, Hans Geiger, Henry Mosley, Ernest Marsden, Owen Chadwick, Charles G. Darwin, George de Hevesy, and Niels Bohr.

Japanese scientists were not unknown at the Cavendish. Three Japanese scientists, T. Noda, T. Kikuchi, and Takeo Shimizu, had already studied there. Shimizu had gained the recognition of the Laboratory's researchers for increasing the cloud chamber's efficiency by mechanically moving the piston with a reciprocating motion and then adding a special camera to "take a great number of photographs of α-ray tracks."[9] Rutherford had also previously accepted a few Japanese physicists into his laboratory at the University of Manchester. There, S. Kinoshita had published a paper on radioactivity in *Proceedings of the Royal Society* and, upon his return to Japan, published two more papers on that subject in *Philosophical Magazine*.[10] However, none of these Japanese physicists had attained prominence or imported the distinctive, highly influential "Cavendish style" of physics research to the Japanese physics community.

Now Nishina's work also was disappointingly less productive than expected. Although he kept abreast of recent research in physics and other branches of science by attending meetings of the Cambridge Philosophical Society,[11] he failed to participate in any important projects and did not produce a single paper during his entire stay at the Cavendish (Figure 2.1). From October 1, 1921, when he entered the Laboratory, to August 15, 1922, Nishina "worked on the Geiger counter and on the distribution of electrons by x-ray scattering using the Geiger counter."[12] According to a letter

FIGURE 2.1 Nishina is the first from the left in the second row in this 1922 annual Cavendish photograph. (Courtesy of the Cavendish Laboratory University of Cambridge.)

he wrote to Bohr in late March of 1923, he had been "counting β-rays excited by γ-rays by means of Geiger's counter" at the Cavendish.[13]

In the traditional Cavendish atmosphere of resolute individuality and independence, Nishina's inexperience no doubt contributed to his lack of productivity. Newcomers were on their own when it came to finding help; the Laboratory's senior staff took no extra pains to assist foreigners or beginners. Certainly, to Japanese physicists like Nishina, the Cavendish Laboratory seemed an interesting — but very strange — place where, as Bohr had indicated a decade earlier, "the state of molecular chaos" prevailed.[14] Although Nishina hoped to expand on Nagaoka's research, he had not been adequately trained to meet this ambitious goal, and he did not receive such training after his arrival at the Cavendish. X-ray spectroscopy was a field to which Cavendish researchers contributed little, as Karl Manne Georg Siegbahn's 1925 book, *The Spectroscopy of X-Rays*, pointed out.[15] Besides the Swede, Karl Siegbahn, the researchers most enthusiastically interested in x-ray spectroscopy were the British father–son team of Henry and Lawrence Bragg; the French Maurice de and Louis de Broglie; the Dutch physicist, Dirk Coster; and the American, Arthur Holly Compton, all of whom were working elsewhere. In the early 1920s, only three Cavendish researchers, A. Müller, P. W. Burbidge, and G. Shearer, published any papers related to x-ray spectroscopy.[16] Rutherford's main concern since 1918 had been investigating the structure of the nucleus by researching the disintegration of light nuclei by bombardment with α-particles. Moreover, although Nishina's study plan required utilizing Geiger counters, this instrument was not of major interest to Cavendish researchers in those years.[17]

Nevertheless, Nishina's one-year long stay at the Cavendish Laboratory was not altogether a waste of time. Although experimentation on x-ray spectroscopy had not yet taken root at the Cavendish, valuable knowledge about x-rays and techniques for scattering had been accumulating there for 20 years.[18] Even Arthur H. Compton, who in 1919 had been elected as a fellow of the National Research Council, chose to spend that academic year using the Cavendish resources to experiment with the scattering and absorption of γ-rays.[19] Nishina was not as well prepared to use the Cavendish's resources as Compton had been, of course, but he did manage to pick up some knowledge about x-ray spectroscopy and techniques for scattering experiments.[20] Also, in the spring of 1922, Nishina first encountered Bohr, an event that soon would lead him to life-altering opportunities.

In late summer of 1922, Nishina left Cambridge and headed for the University of Göttingen in Germany, this time planning to systematically study theoretical physics. Again, he was disappointed. Although he dutifully attended seminars headed by Max Born and David Hilbert, the university's physics program was not yet fully developed.[21] Born's celebrated three-year course on theoretical physics, which was closely related to the experimental training in James Franck's laboratory, and which consisted of "six series of lectures corresponding to the six semesters" had just begun to form.[22] Moreover, as Nishina later remembered, postwar Germany "was very poor indeed," and Nishina, dependent as he was on a fixed income transferred from the Japanese government, found it "very difficult to live."[23] In Germany's chaotic economic situation, inflation was so high that 50,000 marks would buy only a few postage stamps; a handful of stamps amounted to the equivalent of what only a few years before had been the value of a house, and Born was dismayed to actually receive a few postage stamps from a debtor as a mortgage payment.[24]

In November of 1922, as his two years of study abroad were drawing to a close, Nishina received word of his mother's death. This news left him with no reason to return immediately to Japan. Nishina had yet to experience any real success in his research, and physics no longer seemed a tenable career. He was rethinking his plans for the future. Nishina's younger son, Kojiro, provided some insight into Nishina's state of mind at this time:

> What [plans] did he have in mind [for] when he returned to Japan? According to his diary, around the end of his stay in England, he visited an exhibition of physiological instruments and also some instrument workshops in London. This record is in detail and very serious. Perhaps he might have considered producing that kind of instrument after returning to Japan. Interestingly, at the same period, he began to pay attention to science toys. He subscribed to *Games and Toys* and visited the British Association of Toys [sic]. He even considered making wireless control [led] toys and commented that German toys were much better than the English . . . .
>
> Many family members and relatives still believe that during those two years in England and Germany, Father decided that, "Physics is not so interesting to spend my whole life." They argued that he made up his mind "to give up physics and to return to Japan in order to develop and produce science toys. A reason why German science is top in the world is that there is an excellent German science toy industry." Although this argument is not always consistent with his memos or diary, it was anyway a settled opinion among the family . . . . Two wooden boxes of science toys arrived at the family, soon after the letter to his elder brother was delivered.[25]

Despite his doubts about the feasibility of physics research as his future career, on March 25, 1923, before returning to Japan, Nishina sent the following letter to Bohr at the University of Copenhagen:

> To Professor N. Bohr
> Theoretical Physics Dept.
> University of Copenhagen
> Denmark
>
> Dear Sir
>
> You may remember that I was working in the Cavendish Laboratory when you came to Cambridge about a year ago. At that time I was counting β-rays excited by γ-rays by means of Geiger's counter, and had the honour of speaking to you in the laboratory.
>
> I left Cambridge last September and came here for the purpose of learning the German language.
>
> As I spoke to you in Cambridge, I have the great desire of studying in Copenhagen under your guidance, and I should be greatly obliged to you if you could accept me.
>
> As my Institute in Tokio does not allow me to stay in Europe longer than two more terms, I do not know whether it is wise to set up new work. My chief wish is to study

your theory of spectra and atomic constitution in details. But if any one wants assistance in the experiment or the calculation, I should do it with pleasure.

I should esteem it a favour if you would give me the early information in the matter.

I beg to remain

Yours faithfully

Y. Nishina

P.S. I belong to the Institute of Physical and Chemical Research in Tokio,[26] to which Dr. Takamine also does as you know.[27]

Bohr responded by inviting Nishina to study at Copenhagen, to Nishina's surprise and delight. "It is my great pleasure to be able to study in your institute," Nishina wrote back, "and I should like to express my sincere gratitude to you for your kind acceptance of me."[28] On April 10, 1923, Nishina arrived in Copenhagen, where, he wrote to his sister in Japan, he planned "to stay for few months."[29] However, his stay would last for more than 5 years and would prove to be the most fruitful time of his life.

2.3 RESEARCH IN EXPERIMENTAL SUBJECTS AT COPENHAGEN

In 1923, the Institute of Theoretical Physics at the University of Copenhagen had been in existence for only two years (Figure 2.2). With Bohr, the 1922 Nobel Laureate in physics, as its head, and with generous funding by the Danish government, the Carlsberg Foundation, the U.S.A. International Education Board, the Rask-Ørsted

FIGURE 2.2 The Institute of Theoretical Physics at the University of Copenhagen and its director, Niels Bohr. The institute was the center of development of the new quantum mechanics during the 1920s and 1930s. (Courtesy of Niels Bohr Archive.)

Foundation, and other agencies, the Institute quickly emerged as a major physics research center that attracted many talented young scientists from all over the world.[30] During the 1920s and 1930s, Bohr's Copenhagen Institute would wrest the coveted reputation of being the world's most attractive physics center from the Cavendish Laboratory. Proponents of the new quantum mechanics flocked to Copenhagen to exchange ideas, bringing into existence two new terms: the "Copenhagen spirit" and the "Copenhagen interpretation."

From its inception, Bohr's Institute of Theoretical Physics paid almost as much attention to experimentation as it did to theory. In his 1921 inaugural speech, Bohr made it clear that theoretical research on the atom depended on several experimental studies, namely, spectroscopy, ionization by collisions, and radioactivity:

> Among the kinds of scientific work for which the Institute is arranged, spectroscopic research will occupy first place. This is closely connected with the circumstance that by the investigation of the light that matter can be brought to emit we have a means of the greatest value for obtaining information about the composition of matter and of the structure of the atoms of the various elements. Such investigations are made with instruments called spectroscophs [*sic*], and you will have occasion to see several types of these set up in the laboratory rooms.[31]

Bohr's Institute was what Nishina had been searching for, a laboratory where x-ray spectroscopy was of major interest. The importance of spectroscopy in Bohr's Institute during the early 1920s has been nicely summarized by Finn Aaserud:

> By the mid-1920s, Bohr had largely achieved his goal of unifying theory and experiment. During this period, many experimental physicists joined the theoreticians at the institute to make use of the growing collection of apparatus. Among the foreign visitors, James Franck from Germany and Dirk Coster from Holland made particularly significant contributions to spectroscopical work during the first few years. Bohr's Danish colleagues Hans Marius Hansen, who moved into biophysics after the mid-1920s, and Sven Werner also did important experiments in spectroscopy. Jacobsen, who remained an experimentalist at the institute throughout his career, contributed experimentally both in spectroscopy and in radioactivity, the work in radioactivity untypically having little impact on the theoretical work at the institute. Although the discovery of the chemical element hafnium in 1922, which dramatically corroborated Bohr's atomic theory just in time for him to present this finding in his Nobel Prize lecture, is probably the most publicized case of the close interrelationship between theory and experiment at the institute, experimental activity complementing theoretical work went far beyond a single incident.[32]

Under the close supervision of Bohr and other specialists in spectroscopy, Nishina finally could start systematic research on x-ray spectroscopy. Bohr's mentorship was crucial to Nishina's development. Bohr quickly recognized Nishina's talent in experimentation and encouraged him to work harder. When Riken terminated Nishina's financial support in the summer of 1923, Bohr secured fellowships for him from 1924 to 1927 from the Rask-Ørsted Foundation. The foundation was established in 1919 "for the support of Danish science in connection with international research."[33] During the 1920s, it provided an average of three fellowships annually for young foreign scientists, such as Nishina and Wolfgang Pauli, for example, to study at Bohr's Institute.

Nishina worked very hard in Copenhagen. He returned to his lodgings late at night, often climbing over the institute's fence. In October 1923, Bohr's assistant, Hendrik A. Kramers, recorded that "Nishina has left on a little vacation trip to Norway and Sweden together with two other Japanese. We think he has deserved it, and he sends all of us beautiful postcards."[34] Now Bohr thought Nishina might be working too hard. "How could Nishina work so late every day?" Bohr wondered. "Is it because he is working on an experiment? If he is working on theoretical problems, he must not do so."[35] According to Nishina's son, Kojiro:

> [Nishina] wrote in 1925 that "I would like to stay here as long as possible and to study the fundamental principles of physics." His high school classmates worried [about] his strangely long stay in Europe, saying that "Is there anything wrong in Nishina? Does he have a girl friend there? Somebody must go and check whether he is O.K."
>
> For him, the years when he stayed in Europe were the most fruitful as researcher. He once told me that "When I was young and absorbed in studying physics, I once considered not to marry but to concentrate my whole life on research." The period when he stayed in Denmark was certainly those happy years.[36]

While researching x-ray spectroscopy, Nishina enjoyed the good fortune of securing help from two of the period's most distinguished experts on the subject, George de Hevesy and Dirk Coster, both of whom had joined the Institute at Bohr's invitation (Figure 2.3). The Hungarian-born Hevesy had worked on radioactivity and x-ray spectroscopy in various major European institutes, including Rutherford's laboratory in Manchester from 1911 to 1914. He was a good friend of Bohr's ever since Bohr arrived at Rutherford's Manchester laboratory in 1912.[37] The Danish physicist Coster had studied physics under Paul Ehrenfest at Leiden University and x-ray spectroscopy in Karl Siegbahn's laboratory at Lund from 1920 to 1922.[38] At Bohr's request, Coster examined whether Bohr's new atomic theory matched the x-ray spectroscopy data.

In 1922, Hevesy and Coster carried out an experiment to discover the unknown element with the atomic number of 72.[39] According to Bohr's theory, the missing element should not be one of the rare earth materials, but rather a homologue of zirconium [Zi(40)]. Bohr's theoretical prediction was confirmed when Hevesy and Coster discovered the target element in zirconium minerals. They named the new element *hafnium* to commemorate Copenhagen. Although the great moment of discovery occurred before the spring of 1923 when Nishina arrived at Bohr's Institute, Coster greatly influenced Nishina's work on x-ray spectroscopy, and Hevesy became a close friend and generous mentor whom Nishina often consulted not only while he resided in Copenhagen, but also after his return to Japan as well.

Nishina performed his first research at Bohr's Institute with Coster and S. Werner. The three researchers examined the measurement of the absorption-spectra in the L-series of elements from La (57) to Hf (72).[40] These examinations were intended to support Bohr and Coster's 1923 argument that the completion of inner groups of electrons in the atoms was in the neighborhood of the iron, palladium, and platinum groups as well as in the rare earth elements. The researchers carefully measured

FIGURE 2.3 George de Hevesy (left) and Dirk Coster (right). From these two distinguished experts, Nishina learned x-ray spectroscopy and soon became a specialist in that subject. (Courtesy of Niels Bohr Archive.)

the L-absorption spectra of nine elements, namely, La (57), Ce (58), Gd (64), Dy (66), Er (68), Tm (69), Yb (70), Cp (71), and Hf (72). As Coster, Nishina, and Werner reported in the *Philosophical Magazine*, their results "agreed well with the interpolated values . . . used by Bohr and Coster in their paper."[41] Interestingly, the trio submitted this paper on August 13, 1923, just four months after Nishina's arrival at the Institute. Nishina was learning about x-ray spectroscopy very quickly.

Nishina expanded this joint work in his next paper, an individual effort, "On the L-absorption Spectra of Elements from Sn (50) to W (74) and Their Relation to the Atomic Constitution."[42] Although this work aimed to support Bohr's idea of atomic structure, Nishina carefully aimed for "more accurate values of the energy levels" in order to free the interpretation of his results from "the uncertainty involved in the interpolation and arising from the lack of uniformity of the measurements of different authors."[43] Nishina achieved his goals: "A plotting of the square roots of the level values as a function of the atomic number confirm[ed] in a striking way the general conclusions of Bohr and Coster" (Figure 2.4).[44] Moreover, Nishina's experimental results were so accurate that Siegbahn adopted Nishina's findings when he published the English translation of *The Spectroscopy of X-Rays* in 1925.[45] However, Nishina made no attempt to amend Bohr's model, nor did he add any new ideas of his own to the ongoing discussion of x-ray spectroscopy: "The present measurements allow us to examine these anomalies with great accuracy, and to obtain a general support of the conclusions of Bohr and Coster."[46]

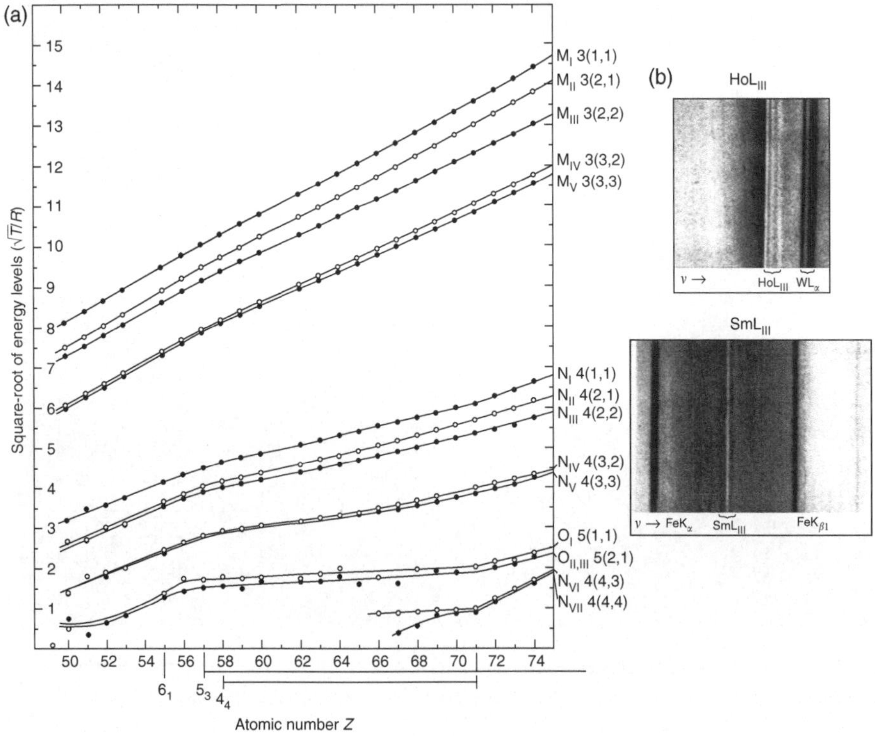

FIGURE 2.4 A diagram ([a] from p. 534) and photographs ([b] from p. 633) from Y. Nishina, "On the L-Absorption Spectra of the Elements from Sn (50) to W (74) and Their Relation to the Atomic Constitution," *Philosophical Magazine*, *49* (1925), 521–537, 633.

In 1925, Coster and Nishina published a joint paper on the quantitative chemical analysis by means of x-ray spectra.[47] In Coster's original method, which he presented in 1923 in *Chemical News*, the element of interest was mixed with an element with an atomic number close to it, and the x-ray spectrum of the mixture was found.[48] In theory, comparing the intensities of the corresponding lines of the two neighboring elements was expected to yield knowledge about the element of interest. In practice, however, the intensities of the two lines were not compared. Instead, the quantity of the element added for comparison was carefully monitored until the two lines became equally intense. The quantity of the added element showed the proportion of the element of interest in the original preparation. Because this so-called "balanced method" was the simplest for separating hafnium from zirconium and other chemicals, it was widely used for the determination of hafnium content in different chemical preparations and minerals. However, Nishina soon discovered that Coster's balanced method was neither accurate nor easily manageable. In a series of letters to Hevesy in the autumn of 1924, Nishina mentioned his dissatisfaction with Coster's method and hinted that he was working toward a better one.[49] Nishina was reticent to publish his concerns about the method in view of his consideration for

Coster's feelings, his respect for their teacher–student relationship, and his Japanese modesty. He revealed some of his thoughts on the matter to Hevesy in a letter dated October 24, 1924:

> As regards the publication on chemical analysis with X-Ray, I am not in position to do it now. It would be criticism of Dr. Coster's work. Therefore unless he publishes something I shall not publish anything. And even after he will have published about the applicability of the method, I do not know whether I publish the material I have got, in my name or not. You know my knowledge on X-Ray spectrograph was given by Dr. Coster.
>
> But the result we [Nishina and Thal Jantzen] have got must be published as soon as possible, otherwise someone may find the same thing and criticize Dr. Coster. The question is how to publish it. Perhaps you and Dr. Coster will communicate about the matter and decide what to do. Meantime I can speak Prof. Bohr about the matter.[50]

Bohr and Hevesy together worked out a compromise: Nishina would write the paper with some help from Coster, but Coster would be the first author.[51]

In the improved method, spectral lines were not measured and directly compared; instead, the already known relative intensities of the elements were used to determine the amount of the element in question within the mixture. Stated differently, Coster and Nishina empirically determined whether the relative intensity of the spectral lines was constant for a definite chemical composition of the preparation used. For example, to determine the amount of hafnium in a mixture of $HfO_2 + ZrO_2$, they simply added Ta_2O_5 until the intensity of the $L_{\alpha 1}$ lines of both hafnium and tantalum became equal. Because the relative concentration of Ta_2O_5:HfO_2 (which showed equal relative intensities of spectral lines) was known to be 2.5:1, the concentration of hafnium in the preparation was determined to be 1/2.5 that of Ta_2O_5. Hevesy and Thal Jantzen at Bohr's Institute demonstrated the reliability of Coster and Nishina's modified method when they used it to determine the hafnium contents in minerals and chemical preparations. The modified method produced very good results that were "in good agreement with those found both by chemical analysis and density measurement."[52] Nishina had become a specialist in x-ray spectroscopy.

Nishina and B. B. Ray further developed the method in the next paper.[53] They emphasized the importance of measuring the relative intensities of x-ray lines to the study of the atom's constitution, and then pointed out some difficulties in making these measurements using the ionization chamber and photographic plate. Nishina and Ray were troubled by "the lack of precise knowledge of the relative sensitiveness of these methods for different wavelengths." To overcome the difficulties, and also to obtain more sensitive curves of the L- and M-series of elements, they covered each of the four faces of the anticathode with sheets of iron, cobalt, nickel, and copper and then turned it so that each of the four faces was exposed to the x-rays. Having made these improvements, they got fair results, which, however, significantly differed from those acquired using optical spectra experimentation. Nishina and Ray closely analyzed the cause of this difference.

FIGURE 2.5 Nishina made many good friends at Bohr's Institute in Copenhagen. He also enjoyed Danish lifestyle, including sawing the lumber. (Courtesy of Niels Bohr Archive.)

The next year, in 1926, Nishina became interested in measuring absorption spectra, for which he enlisted the advice of J. Holtsmark of Trondhjem:

> Recently, I have been very much interested in the measurement of absolute strength of spectral absorption lines and am thinking of applying it to X-Ray region, if possible. As I have not much experience in this matter, I should like to ask you something about Füchtbauer's method. I should be very much obliged to you indeed, if you would let me know the following points.[54]

Nishina also secured help from two Japanese colleagues at Bohr's Institute, Shin'ichi Aoyama and Kojiro Kimura. The three physicists measured the K-absorption spectra of calcium, chlorine, and sulphur in different chemical compounds and found that the absorption frequencies changed according to lattice structure and the mode of chemical binding of the different chemical compounds.[55] In this work, Aoyama and Kimura chiefly carried out "the actual experimental part of the work," while Nishina had taken charge of theoretical interpretation and writing the paper.[56]

This would be the last paper Nishina would write concerning an experimental subject during his stay in Europe. From his arrival at Bohr's Institute in the spring of 1923 until the completion of his joint experiment with Aoyama and Kimura in mid-1927 (Figure 2.6), Nishina had successfully completed several delicate experiments and had matured into a competent experimentalist. Bohr had trusted Nishina's skill and asked Nishina to examine his prediction that an element with an atomic number of 93, 94, or 96 would be chemically similar to uranium (92).[57] A silent film produced in 1925 to advertise the Bohr's Institute included at least two stills that showed Nishina doing experiments (Figure 2.7 below).[58]

2.4 THE NEW QUANTUM MECHANICS AND THE KLEIN–NISHINA FORMULA

In 1927, Nishina told his Japanese colleagues that, after they completed their experiment on the K-absorption spectra of calcium, chlorine, and sulphur in chemical

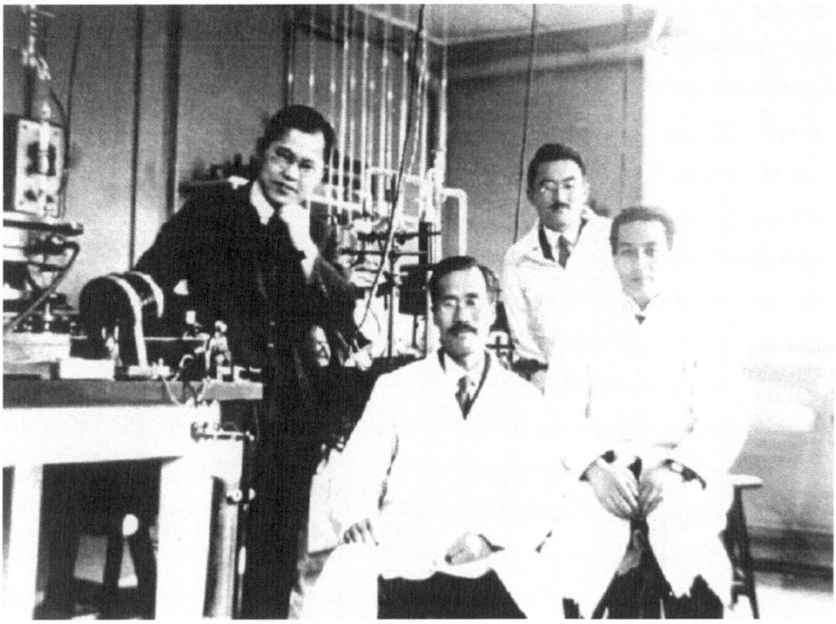

FIGURE 2.6 Nishina and other Japanese scientists at Bohr's Copenhagen Institute in 1927. Standing, Y. Nishina (left) and T. Hori (right); sitting, S. Aoyama (left) and K. Kimura (right). (Courtesy of Niels Bohr Archive.)

compounds, he intended to "stop experimental research for a while and move to theoretical study."[59] The theoretical work Nishina settled on in 1927 was related to the new quantum mechanics that Werner Heisenberg, Wolfgang Pauli, Erwin Schrödinger, Paul Dirac, Oskar Klein, and other young theoretical physicists had been developing since 1925. This was not surprising because Bohr and his Copenhagen Institute had been at the center of this intellectual revolution and, by 1927, Bohr's famous "complementarity" had been born in the unique milieu of the Institute.[60] Bohr had been inviting the principal physicists involved in the development of new quantum mechanics to come together at his institute for work and discussion, and, ultimately, almost every young contributor to the development of new quantum mechanics worked at or visited Bohr's Institute at some point between 1925 and 1935. Kramers, Heisenberg, and Klein served as assistants there; Schrödinger, Pauli, Dirac, Lev Landau, George Gamov, Felix Bloch, John Slater, Friedrich von Weizsäcker, and others working on quantum mechanics visited there for long talks with Bohr or to deliver colloquia. Although Nishina had taken notes on some lectures given by Schrödinger, Friedrich Hund, Dirac, and Heisenberg at Bohr's Institute during 1926–1927,[61] it seems that he had not paid much attention to the new quantum mechanics before his decision to move his work in that direction in mid-1927. However, it does seem natural enough that Nishina would want to study this new development in physics before his return to Japan.

As was customary for Japanese scientists studying abroad, Nishina decided to broaden his travel before returning to Japan. Once again, help came from Bohr,

FIGURE 2.7 Above: Nishina sitting beside experimental apparatus. Below: Bohr (left), Nishina and B. B. Ray (right) standing in front of equipments. Nishina carried out many experiments during his Copenhagen years. (Courtesy of Niels Bohr Archive.)

who arranged for Nishina to follow his ambition to "learn theoretical things from Pauli."[62] Bohr secured 2500 Danish krone in additional financial support from the Rask-Ørsted Foundation to enable Nishina to extend his European stay to study quantum mechanics. Nishina arrived in Paris on August 15, 1927, prepared to learn French and then, in the winter, to study under Pauli at the University of Hamburg in Germany. In early September, Nishina wrote Bohr a letter filled with enthusiasm for studying theoretical physics and with gratitude for Bohr's assistance.[63]

On October 30, Nishina arrived at the University of Hamburg, where he attended Pauli's seminar on theoretical physics from November 8 to February 21 of the following year. In this lecture series, Pauli covered developments in theoretical physics, including the Bohr–Kramers–Slater paper (November 8 lecture), the theory of measurement (December 13 lecture), the spin of electrons (January 10 lecture), Dirac's theory of the electron (February 14 lecture), and the theory of the electrons in metals (February 21 lecture). Pauli's lectures were supplemented by a lecture by Charles G. Darwin on Schrödinger's equation (January 17) and lectures by Otto Stern on thermodynamics and Langmuir's electron theory (January 24 and 30).[64]

Nishina took careful notes during Pauli's seminar, often in Japanese and sometimes in English, as shown in Figure 2.8. His use of Japanese while taking these notes suggests that he was in a hurry to write down what he heard and that he was not yet familiar with the contents of the lectures or the terms used. Interestingly, Nishina took his first notes on the "intensity of Compton's effect," the topic that would eventually bring him fame, while attending Pauli's lecture on Dirac's electron theory on February 14, 1928 (see Figure 2.9).

The University of Hamburg provided an ideal setting for Nishina's study of new development in theoretical physics. Isidor I. Rabi, whom Bohr sent with Nishina to study under Pauli, later remembered the stimulating environment at Hamburg:

> Hamburg actually was the greatest institution in the world for physics at that moment. Hamburg had Pauli; Walter Gordon [of the Klein–Gordon equation]; Wilhelm Lenz, who was in molecular theory, a brilliant man; and most of all, Otto Stern, in experiment. So there quite by accident, and partly against my will, I found myself in this very marvelous place. In addition, there was Ronald Fraser from Scotland, and John Taylor, who was an American. They had both done molecular beams before, and were working now with Stern. Pauli at that time, and this is toward the end of 1927, asked Nishina and me to write a paper with him.[65]

Under Pauli's guidance, Nishina and Rabi published a paper on the dispersion of x-rays in 1928.[66]

After four months of intensive study and work on theoretical topics, Nishina felt much more confident in his ability to explore the new topic of quantum mechanics. By February 19, as he was preparing to return to Copenhagen in early March, he had become deeply involved in theoretical research, as he revealed in a letter to Bohr:

> Dirac's last paper was a great excitement to us here [in Hamburg]. Gordon has worked out the case of Hydrogen atom and obtained the complete Sommerfeld formula with a little different numeration of j. Dirac's letter to Gordon tells that Darwin has obtained the Sommerfeld formula, too. I suppose you have already heard of it.[67]

Immediately upon his return to Copenhagen in the beginning of March, 1928, Nishina began collaborating with Oskar Benjamin Klein to use Dirac's electron theory to calculate the cross section for Compton scattering. Klein already was a well-known theoretician (Figure 2.10).[68] Born in 1894, he was the third child of Gottlieb Klein, the first rabbi in Sweden, and had been a precocious child with a keen interest in the natural sciences, who had enjoyed reading Charles Darwin's *Origin of Species*

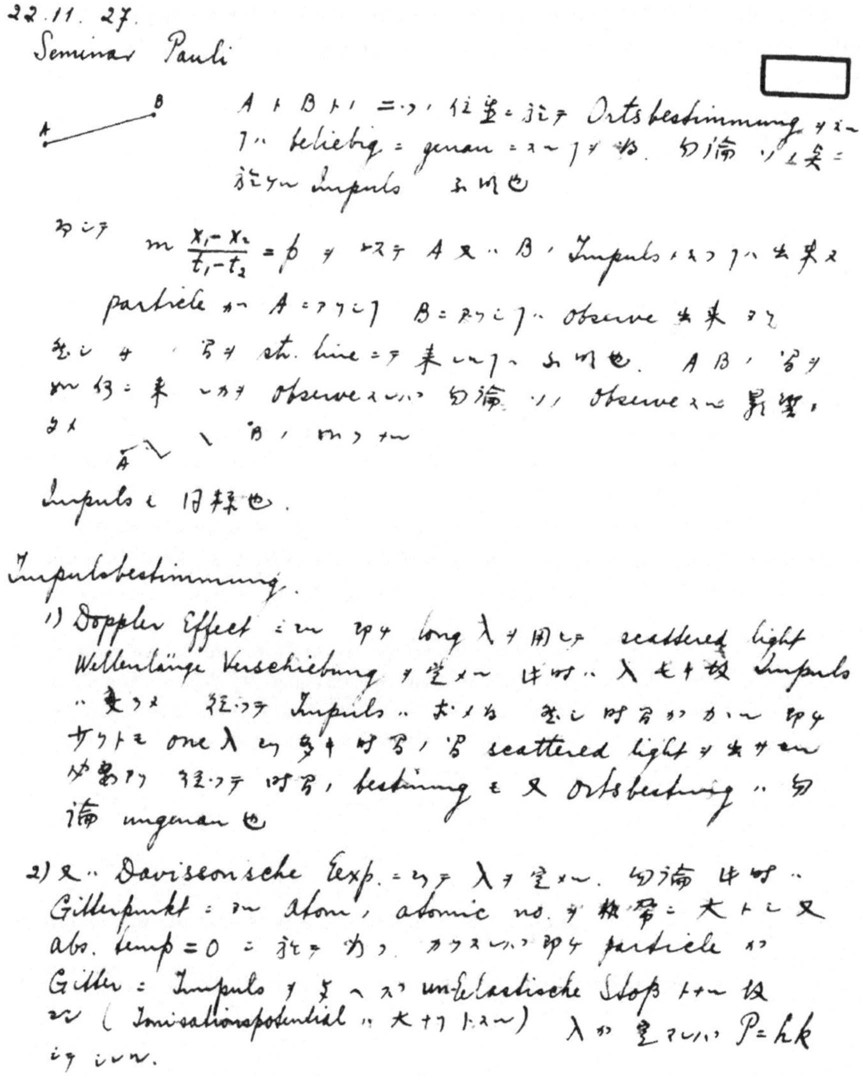

FIGURE 2.8 Nishina's notes on a Pauli seminar lecture, dated November 22, 1927. Nishina wrote these notes mostly in German but sometimes used Japanese and English. (*Source:* Nishina Archive, MSS 103. [Courtesy of the Institute of Physical and Chemical Research.])

as a teenager. At 16, he began working in the Svante Arrhenius' laboratory at the Nobel Institute; at 18, he published his first paper on the solubility of zinc hydroxide. As a university student, Klein continued his work on problems in physical chemistry and produced a few more papers on the dielectric properties of dipolar molecules and electrolytes.

Early in 1918, at the age of 24, Klein wrote to Bohr, asking permission to study quantum physics under his guidance. Bohr welcomed Klein's application, and Klein

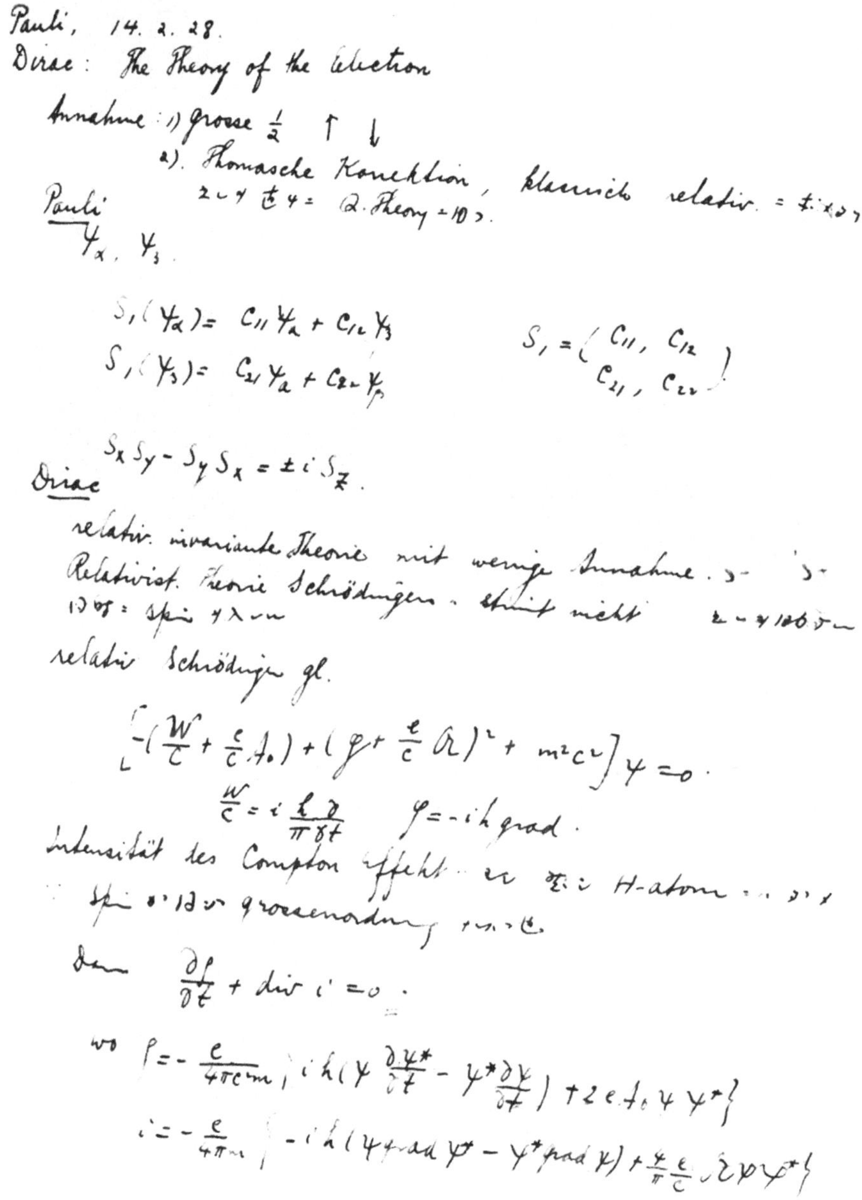
Pauli, 14. 2. 28.
Dirac: The Theory of the electron
Annahme: 1) grosse ½ ↑ ↓
2). Thomasche Korrektion, klassisch relativ.
Pauli
ψ_α, ψ_β.
$S_1(\psi_\alpha) = c_{11}\psi_\alpha + c_{12}\psi_\beta$
$S_1(\psi_\beta) = c_{21}\psi_\alpha + c_{22}\psi_\beta$
$S_1 = \begin{pmatrix} c_{11}, & c_{12} \\ c_{21}, & c_{22} \end{pmatrix}$
Dirac
$S_x S_y - S_y S_x = \pm i S_z$.
relativ. invariante Theorie mit wenige Annahme.
Relativist. Theorie Schrödingers.
relativ Schrödinger gl.
$\left[-\left(\frac{W}{c} + \frac{e}{c}A_0\right) + \left(p + \frac{e}{c}A\right)^2 + m^2c^2\right]\psi = 0.$
$\frac{W}{c} = i\frac{h}{2\pi}\frac{\partial}{\partial t}$ $p = -ih\,\mathrm{grad}$.
Intensität des Compton effekts ... H-atom
$\frac{\partial\rho}{\partial t} + \mathrm{div}\, i = 0$.
wo $\rho = -\frac{e}{4\pi c^2 m}\{ih(\psi\frac{\partial\psi^*}{\partial t} - \psi^*\frac{\partial\psi}{\partial t}) + 2eA_0\psi\psi^*\}$
$i = -\frac{e}{4\pi m}\{-ih(\psi\,\mathrm{grad}\,\psi^* - \psi^*\,\mathrm{grad}\,\psi) + \frac{4\pi e}{c}A\psi\psi^*\}$

FIGURE 2.9 In his notes about Dirac's theory of the electron, dated February 14, 1928, Nishina first noted the "intensity of the Compton effect." (*Source:* Nishina Archive, MSS 103.)

arrived in Copenhagen in May of that year. Between 1923 and 1925, Klein worked at the physics department of the University of Michigan in Ann Arbor, where he developed the five-dimensional theory for the unification of electromagnetic force and gravitation. He visited Copenhagen frequently and, during the 1920s, along

FIGURE 2.10 Paul A. M. Dirac (left) and Oskar Klein (right). Dirac's relativistic theory of electrons of 1928 led Klein and Nishina to work together on Compton scattering. (Courtesy of Niels Bohr Archive.)

with Kramers, collaborated closely with Bohr. In 1925, he returned to Copenhagen and resumed his theoretical work with Bohr. In 1926, he published work on a relativistic version of Schrödinger's wave equation that would become known as the Klein–Gordon equation. Also in that year, Klein became deeply involved in Bohr's work on correspondence and complementarity and, in 1927, together with Pascual Jordan, developed a method of quantization. That year he was appointed Bohr's assistant, with the title of lecturer, a position he would hold until 1930. In early 1928, after Dirac published his famous paper on the relativistic theory of the electron, Bohr sent Klein to meet Dirac at Cambridge.

The accepted explanation for the Klein–Nishina collaboration, as Klein himself related in his essay, is that the collaboration was first suggested by Walter Gordon, who, like Klein, had been working on various theoretical problems since the early 1920s, and that the relation between the two were equal from the beginning:

> When Dirac paid a short visit to Copenhagen in the spring of 1928, he met Klein and Nishina. The three of them were once conferring in the library of the Niels Bohr Institute. Dirac was a man of few words, so when the remark came from Nishina that he had found an error of sign in the then new Dirac paper on the electron, Dirac dryly answered: "But the result is correct." Nishina, in an attempt to be helpful, said: "There must be two mistakes," only to get Dirac's reply that "there must be an even number of mistakes."
>
> The reason behind Nishina's seemingly thorough knowledge of Dirac's paper was that he had been assigned to compute the intensity of the scattered radiation (i.e. the cross

> section) for Compton scattering with the new electron theory. Walter Gordon, during a short visit to Copenhagen somewhat earlier that spring, had suggested that it would be a suitable problem for the young Japanese physicist. He knew Nishina, who had visited Hamburg to learn theory before returning to Japan. When Gordon made that suggestion, Klein immediately agreed, although he himself had intended to attack the same problem. Klein had in fact partly treated Compton scattering the preceding year and Gordon had given it a full treatment, both using semi-classical theory.[69]

However, in a 1963 interview with John L. Heilbron, Léon Rosenfeld, and Thomas S. Kuhn, when Klein was 69, he delivered a slightly different account of his cooperation with Nishina in 1928:

> I think I mentioned to you that when I came back from Dirac [in early 1928], I had thought a little bit about attacking the Compton effect. I had been interested in Compton effect earlier but had never carried it out with the scalar wave equation because Gordon's paper came before I had really begun the real calculation. Then Nishina came to Copenhagen, and he wanted to study theory more closely; he hadn't so very much time before going back to Japan. He had always been interested in theory. He had visited us in the country together with Kronig the summer before, and we had had long talks. He always wanted to have the view on things and followed it in a general way. But then he wanted to do some work also. I think that there was also a curious coincidence here. He was in Hamburg, where he had met Gordon. Gordon had said that perhaps the Compton effect would be something for him. Then he came to Copenhagen, and I had been thinking about the Compton effect. *So Nishina and I decided that he should try it and I should try to guide him. But then the whole became so difficult that I got quite involved in the thing also.*[70]

What prompted Klein to start looking for a formula for Compton scattering in 1928? Since Röntgen's discovery of x-rays in 1895, their nature had been thoroughly studied by many distinguished physicists, including J. J. Thomson, Arnold Sommerfeld, C. T. R. Wilson, W. Henry Bragg, C. Barkla, C. G. Darwin, Henry G. J. Mosely, Peter Debye, Kramers, Louis de Broglie, and Arthur H. Compton. By the beginning of the 1920s, the majority of physicists had concluded that x-rays, which showed all the characteristics of ordinary light — reflection, refraction, diffuse scattering, polarization, diffraction, emission, and absorption spectra — were electromagnetic waves, but with very short wavelengths. However, x-rays also were known to exhibit some behaviors that could not be explained by electromagnetic wave theory. For example, when a beam of soft x-rays was directed toward a metal, the metal emitted electrons, a phenomenon that was termed the photoelectric effect. In 1905, Einstein had suggested that the light (or radiation) might be regarded as discrete bundles of energy of the amount $h\nu$ (h is Planck's constant and ν is the frequency of the light), and that photoelectrons are produced when the light quanta are absorbed by matter. In 1912, Owen W. Richardson and Karl T. Compton successfully demonstrated that the energy of the emitted electrons is proportional to their frequency and that the factor of proportionality is close to the value of h calculated from Planck's radiation formula.[71] However, physicists could not reconcile this particle explanation of photoelectric phenomena with the wave theory of light. J. J. Thomson, in 1925, caricatured the battle between the two competing interpretations as "something like one between a tiger and a shark, each is supreme in its own elements but helpless in that of the other."[72] Arthur H. Compton, in his 1926 book, *X-Rays and Electrons*, admitted that the existence of the photoelectrons ejected by x-rays "is an anomaly when we consider x-rays as electromagnetic waves."[73]

Scattered x-rays presented another puzzle that could not be solved using the electromagnetic theory of light.[74] When x-rays hit the matter of an element with a small atomic number, some of the scattered x-rays shifted to longer wavelengths, a phenomenon first discovered by Barkla. One of the first physicists to investigate this phenomenon thoroughly was Arthur H. Compton (Figure 2.11). After performing a series of careful experiments using monochromatic x-rays, Compton found that the scattered x-rays consisted of two lines: one exactly the same as that of the source of the rays and the other with a somewhat longer wavelength, a phenomenon that soon became known as the Compton effect. The electromagnetic theory of light did not explain this phenomenon because the change of wavelength was known to be independent of the matter collided with even though it varied with the angle between the incident and the scattered rays. Many adhoc theories were put forward to explain the phenomenon, until, in 1923, Compton himself presented the most radical (and successful) explanation by adopting Einstein's idea of light-quantum (or photons).[75] The collision between the incident x-ray and the electron of the matter, Compton suggested, is a collision between two particles: in other words,

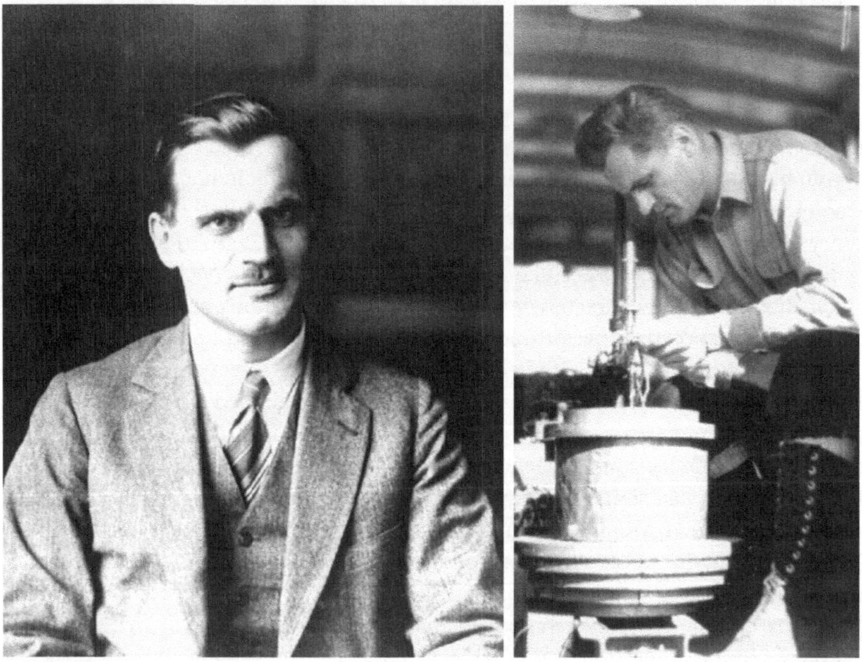

FIGURE 2.11 Arthur Holly Compton doing experiment. Compton's discovery of scattering of light (photons) by electrons, known as "Compton scattering," provided Nishina with an important research topic. (Courtesy of AIP, Emilo Segrè Visual Archives.)

in this case the x-ray is treated as a light-quantum or a photon, with energy $h\nu$ and momentum $h\nu/c$. According to the law of conservation of energy and momentum, the predicted shift in wavelength would depend on the scattering angle θ according to the relation, $\Delta\lambda = (h/mc)(1 - \cos\theta)$. This theory agreed well with the data Compton had carefully collected. Ironically, Compton, who was such a strong proponent of the wave theory of radiation that he chose for his Nobel lecture the topic, "X-Rays as a Branch of Optics," became known as the pioneer of "a new kind of corpuscular theory" of light.[76] Compton was well aware of the effect of his discovery on theoretical explanations of x-rays:

> New scattering phenomena have been observed which are so directly contrary to the usual electrodynamics that we have been compelled to reverse our attitude almost completely. Far from explaining the scattering of X-Rays on the assumption that radiation spreads in all directions as spherical waves, we seem driven by the recent experiments to consider X-Rays as definitely directed quanta of radiant energy.[77]

The Compton effect soon became the focus of interest of physicists. On the one hand, the discovery of this phenomenon revived Einstein's theory of photons and, more generally, the corpuscular theory of light, and thus paved the road to the final "synthesis of matter and light" that occurred within a few years.[78] On the other hand, the discovery provided physicists with a new challenge: how to calculate the Compton effect.[79] In 1926, Compton himself, and also Gregory Breit, Dirac, Gordon, and Schrödinger were in hot pursuit of formulas for measuring the intensity of the scattered radiation by free electrons.[80] Dirac and Gordon, each employing a different method, independently produced the most sensible result by far. However, none of the solutions were entirely satisfactory because none took into account the electron's spin. It was not until 1928 that Dirac proffered the relativistic theory of the electron, which included the electron's spin and thus made possible the development of a formula for calculating the Compton effect.[81] The question now became: Who would tackle this very difficult computation?

For this difficult task, as Gordon perceived, the Klein–Nishina pair would be a dream team. Klein was an expert in all kinds of theoretical tools necessary to solve the problem, and Nishina, who was working hard to master the theoretical side of quantum mechanics, was an expert in calculation. Five years before, in 1923, he had written to Bohr, "If any one wants assistance in the experiment or the calculation, I should do it with pleasure."[82] Moreover, Nishina desperately wanted to accomplish some important theoretical results before his scheduled departure for Japan in the fall. Klein told Heilbron the true nature of this collaboration:

> **Heilbron:** I was wondering in reading over that very long calculation whether or not as a result of that there was some little bit of despair at what physics was coming to, that such an effort was needed to attack a problem which really ....
>
> **Klein:** Oh, I see, but we did the best we could in the short time we had, you see. *Nishina was leaving, so therefore we were both very eager to have a result, and I would have felt very much ashamed if he would have had to leave before we had got any result. So therefore we also used the short-cuts, as I mentioned*

before, and we carried out the calculations as far as we could. We checked very carefully, so that the formula then really came out correctly, and of course we were very happy that it came to such a simple formula . . . . One might have done it perhaps a little shorter by choosing a special coordinate system, but apart from having an electron at rest, I think we didn't specialize the coordinate system with respect to the rays of light.

Heilbron: Since the calculation, of necessity, is so long and involved, I was wondering if there was any feeling that physics has now got to the point that there is . . . .

Klein: Oh, *I think that if I had been alone, I never would have done it.* But then I felt, in some way, that since I had put Nishina on this — or at least contributed to putting him on this — then I was very strongly bound to help to have it carried through. That is, of course, a very great advantage with joint papers; the one drives the other and the other drives the one. *Nishina was very energetic, so that if he had known the principles, he might have done it alone; but he didn't know the principles. He was rather a beginner there. He had read Dirac's paper very carefully, but, I mean, still he was quite a beginner. But I wouldn't have done it alone.*[83]

Klein and Nishina started their work after Easter.[84] Nishina studied Dirac's equation closely, but there was as yet no definite way to solve that equation and progress was disappointing during the spring. "The summer came and we hadn't got very far in it," Klein remembered. "We were uncertain if we should expect just a new proof of the old formula or if there would be a new formula."[85] That summer, Klein and his family stayed at a cottage in Lundeborg on the east coast of the Danish island of Fyn. Klein asked Nishina to work with him there, and Nishina stayed in a nearby pension. Their work became even more intensive, as Klein remembered:

Nishina came with us to the country . . . . Then we were trying very hard on that. I was trying, first, to use the method that I had outlined in that correspondence paper. I had intended to use it here. That was to have the electron waves in a box, to make the states discrete, and then to calculate the transitions between such discrete states. I tried to have the Dirac waves in a box, but one couldn't do that because there were four components there, and one couldn't make them all zero without making the whole zero. Then I tried to have a potential to include them, to have it more physical, and I couldn't do that either . . . . Nishina had to go away, so finally I looked at Gordon's paper and used the way he had done it. Many years after one of my students discovered that that was really only correct for an electron at rest; it wasn't correct generally . . . .

Then there was a lot of such Dirac algebra. I was rather interested because I had looked rather much at these equations, so that [it] came gradually clearer after we had got the principal thing of it. Then there was a lot of algebra to get a definite formula, so we began to fear very much that we would make errors of calculation. We decided to separate while we were doing them, and each of us did them independently. He sat in his room at the pension, I sat at home, and we calculated and compared. Of course, first our things differed. I really got the correct result first, but it was first by only a very little bit — he got it very immediately afterwards. But we checked all those; we did every detail, both of us, on all those calculations. They could be made shorter than we made them, but that was quite a heavy calculation. I think Nishina had found some paper which gave experimental curves, and then we made a theoretical curve. We had not much to draw with, so we did that just a little primitively. But then I think he went to Copenhagen and had it nicely drawn out. So we sent a letter to *Nature* about it, and then we sent the paper.[86]

Manuscripts that show Klein and Nishina's laborious calculations during the summer of 1928 are well preserved in the Nishina archive in Riken. Some years ago, Yuji Yazaki analyzed these manuscripts in detail and published a series of papers in the *Kagakusi Kenkyu*.[87]

On September 15, 1928, *Nature* published a short summary of Klein and Nishina's results entitled, "The Scattering of Light by Free Electrons according to Dirac's New Relativistic Dynamics (dated on August 3, 1928)."[88] On October 30, Klein and Nishina sent the full paper, written in German, but with the same title, to the *Zeitschrift für Physik*, which published it in January of 1929.[89]

As Klein and Nishina pointed out, the formula for measuring the intensity of the scattered radiation, independently developed by both Dirac and Gordon, agreed well with experimental results reflecting the "older form of relativistic quantum mechanics." This formula required modification because Dirac's "new dynamics of the electron" automatically included the effect of the electron's spin. Thus, Klein and Nishina "tried to attack the problem of scattering radiation on free electrons" on the basis of Dirac's relativistic quantum dynamics.[90] In their study, Klein and Nishina considered the incoming radiation to be "a continuous monochromatic wave train," and the pair carefully calculated the intensity of the resulting scattered radiation. In this manner, they arrived at a new formula for the intensity:

$$\begin{aligned} I = I_0 \frac{e^4}{2m^2c^4r^2} \frac{(1+\cos^2\theta)}{\{1+(h\nu/mc^2)\,(1-\cos\theta)\}^3} \\ \times \left(1 + \left(\frac{h\nu}{mc^2}\right)^2 \frac{(1-\cos\theta)^2}{(1+\cos^2\theta)(1+(h\nu/mc^2)\,(1-\cos\theta))}\right) \end{aligned} \tag{2.1}$$

Here I_0 denotes the intensity of the incoming radiation with frequency ν; I denotes the intensity of the scattered radiation at the distance r from the electron; θ denotes the angle between the "the secondary light quantum emitted" and the direction of the incident radiation; e and m denote charge and mass of the electron; c denotes the velocity of light; and h denotes the Planck constant. Klein and Nishina's formula (2.1) differed from Dirac–Gordon's formula by the last factor:

$$\left(1 + \left(\frac{h\nu}{mc^2}\right)^2 \frac{(1-\cos\theta)^2}{(1+\cos^2\theta)(1+(h\nu/mc^2)(1-\cos\theta))}\right)$$

Thus, the Klein–Nishina formula could be reduced into the Dirac–Gordon formula when the photon energy was moderate (or θ was not large), and both became J. J. Thomson's classical formula $I = I_0(e^4/(2m^2c^4r^2))(1+\cos^2\theta)$ when the photon energy was very low (or θ was near zero).

Based on this new formula, Klein and Nishina also calculated the scattering cross section of a substance containing N electrons per unit volume.[91] Klein and Nishina's

1928 publication in *Nature* mentioned the formula's possible application to the "estimation of the wavelengths of the cosmic penetrating radiation." In the upcoming decade, Nishina would make cosmic rays one of the major subjects of his research.

In their original paper, the two authors did not calculate the polarization of the Compton scattered photon. However, Nishina immediately addressed this difficult task on his own and, in the space of just one month, achieved his goal. After Christian Møller of Bohr's Institute pointed out some errors in calculation, Nishina made the necessary corrections and submitted his calculation to *Nature* on September 29, 1928, which published it the following year.[92] Nishina's polarization calculation also was published the following year in the *Zeitschrift für Physik*, in the same volume that included the Klein–Nishina paper on the formula of the intensity of scattering.[93]

Almost immediately, the Klein–Nishina formula was recognized as offering better agreement with existing experimental data on the intensity of Compton scattering than any previous formula. On November 30, 1928, just a month and a half after the publication of the *Nature* paper, Rutherford recognized the success of the formula in his Presidential address to the Royal Society: "Mr. Gray, of the Cavendish Laboratory, who has made a careful examination of existing data on the absorption of gamma-rays, informs me," said Rutherford, "that the evidence as a whole is more in accord with the theory of Klein and Nishina than with the earlier theories of Compton and Dirac."[94] Further experiments by C. Y. Chao (1930) as well as by J. Read and C. C. Lauritsen (1934) confirmed the formula's general validity.[95] From 1930 to 1933, Lise Meitner and H. H. Hupfeld tested the validity of the Klein–Nishina formula in a series of experiments using γ-radiation, and achieved satisfactory overall results.[96] Compton compared several different methods of calculating the intensity of x-ray scattering and declared the Klein–Nishina formula "completely verified" "throughout the wavelength range investigated (Figure 2.12)."[97]

Most interesting to the majority of physicists, however, were the theoretical implications of the Klein–Nishina formula. Many leading physicists considered the formula to be empirical evidence supporting the validity of Dirac's relativistic theory of the electron. It was "the only acceptable early consequence derived from Dirac's relativistic electron theory."[98] Although the Klein–Nishina formula virtually saved Dirac's attractive but troublesome theory, it was Klein who became that theory's principal critic. Just two months after completing the formula, Klein pointed out that the reflection of Dirac's electrons from a reasonably high potential barrier yielded a paradoxical result: the electrons penetrated through and arrived at the other side with negative kinetic energy.[99] Klein's paradox remained a major problem for theoreticians until the discovery of antiparticles. In fact, although Klein and Nishina never explicitly stated so, their formula implied the existence of negative energy states.[100] The possible existence of negative energy states was taken more seriously in 1930 by both Ivar Waller and Igor Tamm: the Klein–Nishina formula could be true only if negative energy states were considered.[101] Physicists later applied the Klein–Nishina formula to the discoveries of the positron, the polarization correlation, and the helicity of the neutrino.

The most impressive aspect of Nishina's collaboration with Klein was Nishina's rapid development as a theoretician. When Nishina began working with Klein in the spring of 1928, he was a learner, a mere *beginner* in theoretical physics; by the end

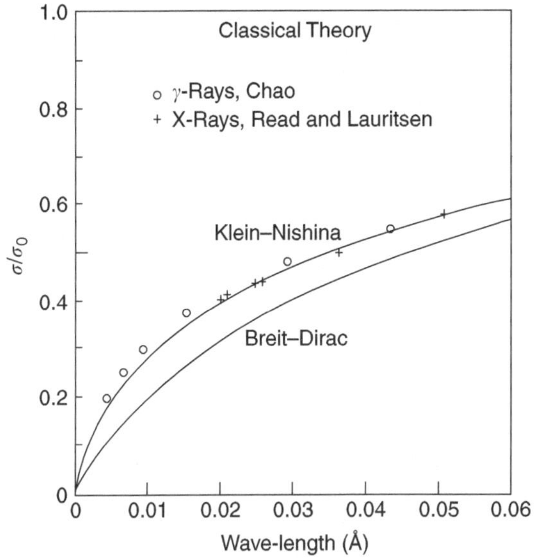

FIGURE 2.12 Various experimental data confirmed the accuracy of Klein–Nishina formula for Compton scattering. (*Source:* Arthur H. Compton and Samuel K. Allison, *X-Rays in Theory and Experiment* [New York: D. Van Nostrand Co., 1935], p. 259.)

of the summer, after completing the Klein–Nishina formula, he was regarded as a *major* player in that field. By September and October, he had no trouble following the contents of lectures given at Bohr's Institute by Gamov, Lothar Nordheim, Douglas R. Hartree, Ralph Fowler, and Nevill Mott, as Nishina's colloquia notes demonstrate.[102]

At the time of the calculation of the Klein–Nishina formula for Compton scattering, Klein was already famous; he had already contributed several important theoretical works to the field of quantum mechanics. Therefore, in terms of fame and recognition, Nishina was the principal beneficiary of the collaboration. From an historical point of view, it might be interesting to know *how much* Klein and Nishina *each* contributed to the discovery of their formula. However interesting, the answer to this question is not important. As a raw beginner in the field of theoretical physics, Nishina *could not* have done it alone, and, as Klein stated 35 years after the discovery, he "*would not* have done it alone."[103] Without this historic collaboration, the formula would not have seen the light of day, at least at that early date.

2.5 A TRULY ACCOMPLISHED RESEARCHER

With the calculations of the Klein–Nishina formula and of the polarization of the scattered radiation, Nishina's more than seven-year stay in Europe finally ended. In late October of 1928, he left Copenhagen for Paris; on the last day of that month, he boarded a ship for the United States. He disembarked in New York City on November 12 and began a rapid coast-to-coast journey that included tours of some important educational institutions, including Arthur Compton's laboratory at the University of Chicago and the Mount Wilson Observatory. On this trip, Nishina reportedly asked American physicists the required expenditure to support travel to Japan by "famous scientists from Europe," presumably Bohr and other European luminaries in the field of physics.[104] On December 5, 1928, he boarded a ship in San Francisco, crossed the Pacific Ocean, and arrived in Kobe on the 21st. His remarkable overseas study finally completed. Until 1949, he would not leave his country again.

What did Nishina accomplish during his long stay in Europe? First, he matured into a competent physicist with excellent experimental skills and considerable knowledge and experience in theory, especially that of the new quantum mechanics. He became a top specialist in x-ray spectroscopy whose experimental work and data were recognized by leading physicists, including Siegbahn, who employed Nishina's data in his 1925 book on x-ray spectroscopy, and Arthur H. Compton, who mentioned Nishina's three experimental works in his 1935 book, *X-Rays in Theory and Experiment*.[105] Nishina produced outstanding results in theory both in collaboration with Klein and working independently on the polarization of Compton scattering. He became the first Japanese physicist whose work made him known to most leading Western physicists. In short, Nishina became a truly accomplished researcher in the new field of quantum mechanics, a very rare achievement, particularly for a Japanese.

Second, during his long stay at Bohr's Institute in Copenhagen, Nishina learned how to manage a large, modern, physics institute. Bohr was the mentor who, perhaps unintentionally, taught Nishina the role of institute director: how to get theoreticians and experimentalists to work together under the same roof and how to establish and maintain a creative environment that stimulated researchers to work hard and independently. The unique Copenhagen atmosphere that Nishina carried with him would contribute greatly to the success of his own laboratory in Riken during the next decade.[106] Although Nishina was one of seven Japanese physicists who worked at Bohr's Institute during the 1920s, he was the only one of these who truly understood the unique Copenhagen atmosphere.[107] In Japan, he would cleverly and successfully adopt this unique milieu to Japanese culture, producing new generations of Japanese physicists.

Third, during his long stay in Europe, Nishina established friendships with Bohr and other leading European physicists. Rutherford, Bohr, Heisenberg, Pauli, Dirac, Hevesy, Coster, and Klein were just few names on the list of his European friends. His links with European luminaries in physics would prove helpful in promoting his research and teaching career in Japan. His European friends kept him informed about the most recent developments in physics and helped him acquire experimental materials from Europe. For example, Hevesy made it possible for him to purchase such rare elements as boron, beryllium, and scandium.[108] More visibly and probably more importantly, Nishina was able to sponsor visits to Japan by some of his esteemed European friends; he planned and carried out a visit by Heisenberg and Dirac in 1929, a visit by Hevesy in 1931, and a visit by Bohr in 1937. The information they brought and the contacts they made possible contributed greatly to the rapid development of physics in Japan during the 1930s.

Despite Nishina's remarkable success during his seven years abroad, when Nishina returned to Japan, no Japanese — including Nishina himself — fully recognized his accomplishments in Europe and his potential for contributing to the study of physics in Japan. Nishina's apology to Bohr for "not accomplishing anything valuable in physics during this long stay" must not be dismissed as typical Japanese modesty. Fortunately, a few somewhat peculiar conditions in Japan around 1930 led Nishina to realize his full potential.

NOTES

1 R. Kubo, "Yoshio Nishina, the Pioneer of Modern Physics in Japan," in M. Suzuki and R. Kubo (eds.), *Evolutionary Trends in the Physical Sciences*, pp. 3–11 on pp. 3–4.

2 N. Bohr to Y. Nishina (January 26, 1934), in *Y Nishina's Correspondence with N. Bohr and Copenhageners, 1928–1949* (Tokyo: Nishina Memorial Foundation, 1984), pp. 31–33 on p. 31.

3 Y. Nishina to N. Bohr (October 21, 1928), in *Y. Nishina's Letters to N. Bohr, G. Hevesy and Others, 1923–1928* (Tokyo: Nishina Memorial Foundation, 1985), pp. 23–24 on p. 23.

4 Kojiro Nishina, "My Father's Overseas Study," in Hidehiko Tamaki and Hiroshi Ezawa (eds.), *Nishina Yoshio* [in Japanese] (Tokyo: Misuzu Shobo, 1991), pp. 266–272 on pp. 266–267.

5 "An Institute of Physical and Chemical Research for Japan," *Nature, 102* (1918), 294–295.

6 K. Nishina, "My Father's Overseas Study," in *Nishina Yoshio*, p. 268.

7 "Nagaoka Laboratory," in "Summary of the Past Activities of the Institute," *SP*, *34* (1938), 1763–1892 on 1826–1833.

8 H. Nagaoka to E. Rutherford (June 20, 1919), E. Rutherford Archive, Cambridge University Library, MSS ADD 7653 N2.

9 Takeo Shimizu, "A Preliminary Note on Branched α-Ray Tracks," *Proceedings of the Royal Society*, *99* (1921), 432.

10 S. Kinoshita, "The Photographic Action of the α-Particles Emitted from Radioactive Substances," *Proceedings of the Royal Society*, *83* (1910), 432–453; S. Kinoshita, S. Nishikawa, and S. Ono, "On the Amount of the Radioactive Products Present in the Atmosphere," *Philosophical Magazine*, *22* (1911), 821–840; S. Kinoshita and H. Ikeuti, "The Tracks of the α Particles in Sensitive Photographic Films," *Philosophical Magazine*, *29* (1915), 420–425.

11 *Proceedings of the Cambridge Philosophical Society*, *21* (1922–1923), 290. At the 1921 annual meeting, held on October 31, Nishina and eight other new members were elected associates of the society.

12 H. Ezawa and H. Takeuchi, "The Chronology of Yoshio Nishina [in Japanese]," in *Nishina Yoshio*, pp. 273–300 on p. 274.

13 Y. Nishina to N. Bohr (March 25, 1923), in *Y. Nishina's Letters to N. Bohr, G. Hevesy and Others, 1923–1928*, p. 1.

14 S. Rozental (ed.), *Niels Bohr: His Life and Works as seen by His Friends and Colleagues* (Amsterdam: North-Holland Personal Library, 1967), p. 41.

15 Karl M. G. Siegbahn, *The Spectroscopy of X-Rays*, translated by George A. Lindsay (London: Oxford University Press, 1925), pp. 271–280. The first German edition was published in 1923, and the author "revised several chapters so as to incorporate the results of recent researches" for the 1925 English edition.

16 Alex Müller, "On an X-Ray Bulb with a Liquid Mercury Anticathode, and on Wave-Length Measurements of the L-Spectrum of Mercury," *Philosophical Magazine, 42* (1921), 419–427; P. W. Burbidge, "The Absorption of the K X-Rays of Silver in

Gases and Gaseous Mixtures," *Philosophical Magazine, 43* (1922), 381–389; "Note on the Absorption of Narrow X-Ray Beams," *Philosophical Magazine, 43* (1922), 389–392; G. Shearer, "The Emission of Electrons by X-Ray," *Philosophical Magazine, 44* (1922), 793–808.

17 J. G. Crowther claimed that "No Geiger counters were used in the Cavendish until about 1930." This cannot be correct because some Cavendish researchers are known to have produced quantitative results using a Geiger counter during the 1920s. Crowther's statement, however, does demonstrate that the Geiger counter was not a popular instrument at the Cavendish at that time. See J. G. Crowther, *The Cavendish Laboratory, 1874–1974* (New York: Science History Publications, 1974), p. 214.

18 See Dong-Won Kim, *Leadership and Creativity: A History of the Cavendish Laboratory, 1871–1919* (Dordrecht, Netherlands: Kluwer Academic, 2002), Chapter 5.

19 Roger H. Stuewer, *The Compton Effect: Turning Point in Physics* (New York: Science History, 1975), Chapter 4.

20 G. Ekspong, "Oskar Klein and Yoshio Nishina," in Suzuki and Kubo (eds.), *Evolutionary Trends in the Physical Sciences*, pp. 25–34 on p. 25.

21 Kenkichiro Koizumi, "Nishina Yoshio during the Overseas Study in Europe [in Japanese]," *Shizen* (November, 1976), 58–67 on 62.

22 Max Born, *My Life: Recollections of a Nobel Laureate* (London: Taylor & Francis, 1978), pp. 210–211.

23 Y. Nishina, "What I Have Read" in *Atomic Power and I*, p. 228.

24 Born, *My Life*, p. 202.

25 K. Nishina, "My Father's Overseas Study," in *Nishina Yoshio*, pp. 267–270.

26 Riken had promoted Nishina from Associate Fellow to Full Fellow shortly after he entered the Cavendish.

27 Y. Nishina to N. Bohr (March 25, 1923), in *Y. Nishina's Letters to N. Bohr, G. Hevesy and Others*, p. 1.

28 Y. Nishina to N. Bohr (April 4, 1923), ibid., p. 2.

29 K. Nishina, "My Father's Overseas Study," in *Nishina Yoshio*, p. 270.

30 For a history of the Institute, see "NBI History" (http://www.nbi.dk/nbi-history.html); Finn Aaserud, *Redirecting Science: Niels Bohr, Philanthropy and the Rise of Nuclear Physics* (Cambridge: Cambridge University Press, 1990); and P. Robertson, *The Early Years: The Niels Bohr Institute, 1921–1930* (Copenhagen: Akademisk Forlag, 1979).

31 Niels Bohr, "Dedication of the Institute for Theoretical Physics (March 3, 1921)," in Léon Rosenfeld (general editor), *Niels Bohr Collected Works, Vol. 3: The Correspondence Principle (1918–1923)*, edited by J. Rud Nielsen (Amsterdam: North-Holland, 1976), pp. 293–301 (translation) on p. 296.

32 Aaserud, *Redirecting Science*, p. 27.

33 Ibid., pp. 25–26.

34 Letter from H. A. Kramers to N. Bohr (October 11, 1923), in *Niels Bohr Collected Works, Vol. 3: The Correspondence Principle (1918–1923)*, p. 662.

35 Kojiro Kimura, "Dr. Nishina at Copenhagen [in Japanese]," in *Nishina Yoshio*, pp. 33–45 on pp. 41–42.

36 K. Nishina, "My Father's Overseas Study," in *Nishina Yoshio*, p. 270.

37 For more about the life and work of George de Hevesy, see his autobiographical essay, "A Scientific Career," in George de Hevesy (ed.), *Adventures in Radioisotope Research: The Collected Papers of George Hevesy*, 2 volumes (New York: Pergamon, 1962), Vol. 1, pp. 11–30; H. Levi, *George de Hevesy: Life and Work* (Copenhagen: Rhodos, 1985); Frence Szabadvary, "Hevesy, George," *DSB*, Vol. 6, 365–367.

38 J. A. Prins, "Coster, Dirk," *DSB*, Vol. 3, 429–430.

39 For information about the discovery of hafnium and its relationship to Bohr's theory of the periodic system, see Helge Kragh, "The Theory of the Periodic System," in A. P. French and P. J. Kennedy (eds.), *Niels Bohr: A Centenary Volume* (Cambridge, MA: Harvard University Press, 1985), pp. 50–67.
40 D. Coster, Y. Nishina, and S. Werner, "Röntgenspektroskopie. Über die Absorptionspektren in der L-Serie der Elemente La (57) bis Hf (72)," *Zeitschrift für Physik, 18* (1923), 207–215.
41 Ibid., 211.
42 Y. Nishina, "On the L-Absorption Spectra of the Elements from Sn (50) to W (74) and Their Relation to the Atomic Constitution," *Philosophical Magazine, 49* (1925), 521–537.
43 Ibid., 521.
44 Ibid., 522.
45 K. Siegbahn, *The Spectroscopy of X-Rays*, pp. 184–187. In his preface (p. vi), Siegbahn acknowledged, "Table 37 and 38 on the energy levels [were] corrected with the help of new measurements kindly put at my disposal by Mr. Nishina."
46 Y. Nishina, "On the L-Absorption Spectra," 535.
47 D. Coster and Y. Nishina, "On the Quantitative Chemical Analysis by means of X-Ray Spectrum," *Chemical News, 130* (1925), 149–152.
48 D. Coster, "X-Ray Spectroscopy as a Means of Qualitative and Quantitative Chemical Analysis," *Chemical News, 127* (1923), 65–70.
49 Y. Nishina to G. Hevesy (October 13, 1924), (October 24, 1924), (November 18, 1924), and (November 29, 1924), in *Y. Nishina's Letters to N. Bohr, G. Hevesy and Others*, pp. 3–4, 4–5, 6–7, and 8–9, respectively.
50 Y. Nishina to G. Hevesy (October 24, 1924), ibid., p. 5.
51 Kojiro Kimura, who had worked at Bohr's institute at the time the paper was published, argued that the experiment was carried out by Nishina alone. See Kimura, "Dr. Nishina in Copenhagen," in *Nishina Yoshio*, p. 39.
52 Coster and Nishina, "On the Quantitative Chemical Analysis by Means of X-Ray Spectrum," 152.
53 Y. Nishina and B. B. Ray, "Relative Intensity of X-Ray Lines," *Nature, 117* (1926), 120–121.
54 Y. Nishina to J. Holtsmark (June 6, 1926), Nishina MSS 336. Holtsmark replied with considerable technical advice [J. Holtsmark to Y. Nishina (June 9, 1926), Nishina MSS 337].
55 S. Aoyama, K. Kimura, and Y. Nishina, "Die Abhängigkeit der Röntgenabsorptionsspecktren von der chemischen Bindung," *Zeitschrift für Physik, 44* (1927), 810–833.
56 Y. Nishina to Bøggild (November 11, 1927), in *Y. Nishina's Letters to N. Bohr, G. Hevesy and Others*, p. 20; Kimura, "Dr. Nishina in Copenhagen," in *Nishina Yoshio*, p. 40. Nishina also corresponded with W. Kuhn for these experiments. See W. Kuhn to Nishina (August 8, 1926), Nishina MSS 338; (October 25, 1926), MSS 339; and (December 14, 1926), MSS 340.
57 Kimura, "Dr. Nishina in Copenhagen," in *Nishina Yoshio*, pp. 43–44. Bohr, Hevesy, and Nishina all expected that a higher atomic number of 94 or 96 would be found and, in late 1924, Nishina wrote to Hevesy seeking his view about whether A. A. Russel's recent work in *Nature* supported their shared opinion. See Y. Nishina to G. Hevesy (November 11, 1924) in *Nishina's Letters to N. Bohr, G. Hevesy and Others*, p. 7.
58 Robertson, *The Early Years*, p. 96.
59 Kimura, "Dr. Nishina in Copenhagen," in *Nishina Yoshio*, p. 40.
60 For more about the scientific activities and unique milieu of Bohr's Institute during this period, see Robertson, *The Early Years*; Ruth E. Moore, *Niels Bohr: The Man,*

His Science, & the World They Changed (Cambridge, MA: MIT Press, 1985); Abraham Pais, *Niels Bohr's Time: In Physics, Philosophy, and Polity* (Oxford: Clarendon Press, 1991); F. Aaserud, *Redirecting Science*, Chapter 2; French and Kennedy, *Niels Bohr*, Parts III and IV.

61 Hiroshi Ezawa, "The Apostle Who Witnessed the Genesis [in Japanese]," *Butsuri*, *45* (1990), 744–751 on 745–748.

62 Y. Nishina to N. Bohr (August 16, 1927), in *Y. Nishina's Letters to N. Bohr, G. Hevesy and Others*, pp. 15–16 on p. 16.

63 Y. Nishina to N. Bohr (September 1, 1927), ibid., pp. 16–17.

64 Nishina Archive, MSS 103.

65 I. I. Rabi, "Stories from the Early Days of Quantum Mechanics," transcribed and edited by R. Fraser Code, *Physics Today* (August, 2006), 36–41 on 38.

66 Y. Nishina and I. I. Rabi, "Der wahre Absorptionskoeffizient der Röntgenstrahlen nach der Quantentheorie," *Verhandlungen der Deutschen Physikalischen Gesellschaft*, *9* (1928), 6–9.

67 Y. Nishina to N. Bohr (February 19, 1928), in *Nishina's Letters to N. Bohr, G. Hevesy and Others*, p. 22.

68 For Klein's life and works, see Ulf Lindström (ed.), *Proceedings of the Symposium of The Oskar Klein Centenary, 19–21 September 1994* (Singapore: World Scientific, 1995), especially the article by A. Pais, pp. 1–22; Gösta Ekspong (ed.), *The Oskar Klein Memorial Lectures*, 2 volumes (Singapore: World Scientific, 1994); K. Meyenn and M. Baig, "Klein, Oskar Benjamin," *DSB*, Vol. 17, pp. 480–484.

69 Gösta Ekspong, "The Klein–Nishina Formula," in G. Ekspong (ed.), *The Oskar Klein Memorial Lectures, Vol. 2: Lectures by Hans A. Bethe and Alan H. Guth with translated reprints by Oskar Klein* (Singapore: World Scientific, 1994), pp. 97–112 on pp. 97–98. For Klein's own memoir, see O. Klein, "Research Work," in *Nishina Yoshio*, pp. 93–97.

70 Oskar Klein interview with John L. Heilbron, Léon Rosenfeld and Thomas S. Kuhn, AIP MSS OH 256, session 4 (February 28, 1963), 16. Italics are added.

71 O. W. Richardson and Karl T. Compton, "The Photoelectric Effect," *Philosophical Magazine*, *24* (1912), 575–594.

72 J. J. Thomson, *The Structure of Light: The Fison Memorial Lecture 1925* (Cambridge: Cambridge University Press, 1925), p. 15.

73 A. H. Compton, *X-Rays and Electrons* (New York: D. Van Nostrand Company, 1926), p. 224.

74 See Roger H. Stuewer, *The Compton Effect*.

75 Arthur Holly Compton, "A Quantum Theory of the Scattering of X-Rays by Light Elements," *Physical Review*, *21* (1923), 483–502.

76 Karl M. G. Siegbahn, "Presentation Speech," in *Nobel Lectures, Physics 1922–1941* (Singapore: World Scientific, 1998), p. 170.

77 A. Compton, *X-Rays and Electrons*, p. 260.

78 Bruce R. Wheaton, *The Tiger and the Shark: Empirical Roots of Wave-Particle Dualism* (Cambridge: Cambridge University Press, 1983), Chapter 10.

79 For physicists' attempts to calculate Compton scattering, see Laurie M. Brown, "The Compton Effect as a Path to QED," *Studies in History and Philosophy of Modern Physics*, *33B* (2002), 211–249.

80 Paul A. M. Dirac, "Relativity Quantum Mechanics with an Application to Compton Scattering," *Proceedings of the Royal Society*, *111* (1926), 405–423; W. Gordon, "Der Comptoneffekt nach der Schrödingerschen Theorie," *Zeitschrift für Physik*, *40* (1926), 117–133. For more about attempts to develop the formula before the Klein–Nishina

collaboration, see Arthur H. Compton and Samuel K. Allison, *X-Rays in Theory and Experiment* (New York: D. Van Nostrand Company, 1935), pp. 231–236 and Ekspong, "The Klein–Nishina Formula," pp. 99–104.

81 Paul A. M. Dirac, "The Quantum Theory of the Electron," *Proceedings of the Royal Society, 117* (1928), 610–624.

82 Y. Nishina to N. Bohr (March 25, 1923), in *Y. Nishina's Letters to N. Bohr, G. Hevesy and Others, 1923–1928*, p. 1.

83 O. Klein interview with Heilbron, Rosenfeld and Kuhn (February 28, 1963), p. 19. Italics are added.

84 For more details about the Klein–Nishina collaboration, see O. Klein interview with Heilbron et al. and "Research Work," in *Nishina Yoshio*, pp. 93–97.

85 O. Klein interview with Heilbron et al., p. 16.

86 Ibid., pp. 16–17.

87 Y. Yazaki, "The Process of Developing [the] Klein–Nishina Formula (I): A Study with the Manuscripts in Nishina Archive in Riken [in Japanese]," *Kagakusi Kenkyu, II, 31* (1992), 81–91; "The Process of Developing Klein–Nishina Formula (II) [in Japanese]," *Kagakusi Kenkyu, II, 31* (1922), 129–137.

88 O. Klein and Y. Nishina, "The Scattering of Light by Free Electrons according to Dirac's New Relativistic Dynamics," *Nature, 122* (1928), 398–399.

89 O. Klein und Y. Nishina, "Über die Streuung von Strahlung durch freie Elektronen nach der neuen relativistischen Quantendynamik von Dirac," *Zeitschrift für Physik, 52* (1929), 853–868. This paper was later translated into English by G. Ekspong and added in *The Oskar Klein Memorial Lectures*, Vol. 2, pp. 113–129.

90 Ekspong, *The Oskar Klein Memorial Lectures*, Vol. 2, p. 113 (English translation).

91 The result was

$$s = \frac{2\pi N e^4}{m^2 c^4}\left\{\frac{1+\alpha}{\alpha^2}\left[\frac{2(1+\alpha)}{1+2\alpha} - \frac{1}{\alpha}\log(1+2\alpha)\right] + \frac{1}{2\alpha}\log(1+2\alpha) - \frac{1+3\alpha}{(1+2\alpha)^2}\right\},$$

where $\alpha = h\nu/mc^2$ and N is the number of electrons.

92 Y. Nishina, "Polarisation of Compton Scattering according to Dirac's New Relativistic Dynamics," *Nature, 123* (1929), 349. The final formula is

$$I = \frac{e^8}{2m^4c^8r^2r'^2}\frac{I_0}{(1+2\alpha)^3}\left\{\sin^2\theta + \frac{\alpha^2(2+4\alpha+3\alpha^2)}{2(1+\alpha)^2(1+2\alpha)}\right\},$$

where $\alpha = h\nu/mc^2$, I_0 is the incident beam of unpolarized beam with the frequency ν, r the distance between the original collision and the second collision (y-axis), r' the distance between the second collision and the point where the measurement is made, θ the angle in the second collision, e and m the charge and mass of the electron, c the velocity of light, and h the Planck constant.

93 Y. Nishina, "Die Polarisation der Comptonstreuung nach der Diracschen Theorie des Elektrons," *Zeitschrift für Physik, 52* (1929), 869–877.

94 Ernest Rutherford, "Presidential Address," *Proceedings of the Royal Society, 122* (1929), 1–23 on 15.

95 C. Y. Chao, "Scattering of Hard γ-Rays," *Physical Review, 36* (1930), 1519–1522; J. Read and C. C. Lauritsen, "An Investigation of the Klein–Nishina Formula for X-Ray Scattering, in the Wave-Length Region 50 to 20 X-Units," *Physical Review, 45* (1934), 433–436.

96 For more about Meitner's research on the Klein–Nishina formula, see Ruth L. Sime, *Lise Meitner: A Life in Physics* (Berkeley, CA: University of California Press, 1996), pp. 120–125.

97 Arthur H. Compton and Samuel K. Allison, *X-Rays in Theory and Experiment* (New York: D. Van Nostrand Co., 1935), p. 259. This is the retitled second edition of Compton's *X-Rays and Electrons.*

98 L. M. Brown, A. Pais, and Brian Pippard (eds.), *Twentieth Century Physics*, 3 volumes (Bristol: Institute of Physics, 1995), Vol. 1, p. 218.

99 O. Klein, "Die Reflexion von Elektronen an einem Potentialsprung nach der relativistischen Dynamik von Dirac," *Zeitschrift für Physik, 53* (1929), 157–165.

100 For further discussion about negative energy in Dirac's theory, see Helge Kragh, *Dirac: A Scientific Biography* (Cambridge: Cambridge University Press, 1990), Chapter 5.

101 I. Waller, "Die Streuung von Strahlung durch gebundene und freie Elektronen nach der Diracschen relativistischen Mechanik," *Zeitschrift für Physik, 61* (1930), 837–851; I. Tamm, "Über die Wechselwirkung der freien Elektronen mit der Strahlung nach der Diracschen Theorie des Elektrons und nach der Quantenelektrodynamik," *Zeitschrift für Physik, 62* (1930), 545–568.

102 In the autumn of 1928 at Bohr's Institute, Nishina took notes on lectures by the following physicists: Gamov (September 3, 1928), Nordheim (September 7, 1928), Hartree (September 12, 1928), Fowler (September 29, 1928), and Mott (October 28, 1928) [Nishina Archive MSS 96]. Nishina's notes on these lectures were shorter than those he took in Hamburg.

103 O. Klein interview with Heilbron et al., p. 19.

104 Hidehiko Tamaki, "The Role of Dr. Nishina [in Japanese]," *Butsuri, 45* (1990), 755–758 on 756.

105 Siegbahn, *The Spectroscopy of X-Rays*, pp. 184–187; Compton and Allison, *X-Rays in Theory and Experiment*, pp. 667, 668, and 794.

106 For the unique Copenhagen atmosphere during the 1920s and 1930s, see Aaserud, *Redirecting Science*, pp. 6–7.

107 The other six were Shinichi Aoyama, Kenjiro Kimura, Yoshikatsu Sugiura, Toshio Takamine, Mitsuharu Fukuda, and Takeo Hori. A few more Japanese physicists briefly visited Bohr's Institute during the 1920s.

108 For more about Nishina's purchase of boron, beryllium, and scandium through Hevesy in 1933, see *G. Hevesy–Y. Nishina Correspondence, 1928–1949*, pp. 7–11. Many letters to Nishina from Bohr and Hevesy included news of recent research in Europe.

3 Preacher of the New Quantum Mechanics

> The myth that new physics in Japan started with the return of Nishina on December 10, 1928 persists . . . even today. In this article, I will point out that several attempts to introduce new quantum mechanics into Japan had preceded his return, and that Nishina became the ideal man to plant the seed of new physics in Japan precisely because the soil had been already well prepared by other Japanese physicists.
>
> **Atsushi Katsuki**[1]

3.1 THE INTRODUCTION OF THE NEW QUANTUM MECHANICS TO JAPAN DURING THE 1920s

The end of World War I in 1918 transformed the Japanese physics community in many ways. First of all, the Great War taught the Japanese that basic science research could improve not only science, medicine, and engineering but also contribute to a nation's economic development and national security. In 1917, to foster basic research in physics and chemistry specifically, Japan founded the Institute of Physical and Chemical Research (Riken). In 1920, Japan demonstrated its new commitment to basic research and the "growing independence of Japanese science" by establishing a National Research Council.[2]

Japan's younger physicists, in particular, revised their traditional views in response to the rapidly changing, post-1918 perspectives in physics. They embraced a deeper understanding of atomic structure, the discovery of the dual nature of light, attempts to validate Einstein's theory of general relativity, and the emergence of the new quantum mechanics. Throughout the 1920s, Japan's social and cultural milieu influenced the country's younger generation of physicists to turn their attention away from applied or experimental subjects and toward more theoretical ones.

For Japanese physicists, the first truly memorable event to occur on Japanese soil was Albert Einstein's six-week visit to Japan in 1922.[3] Einstein's Nobel Prize in physics had been announced during his voyage to Japan, and even ordinary Japanese citizens welcomed the world's most renowned scientist and his wife with particular enthusiasm. "Each of lecture halls was full to overflowing," when Einstein delivered lectures in Tokyo, Sendai, Nagoya, Kyoto, Osaka, Kobe, and Fukuoka.[4] At his first lecture at Keio University in Tokyo, on November 19, about 2000 people listened to him speak for three hours on the theories of special and general relativity.[5] Japanese reporters recorded Einstein's observations and comments on the Japanese people, its culture and its environment for readers. His appreciation of Japanese Noh

FIGURE 3.1 Einstein lecturing at Tokyo Imperial University. His lectures on physics topics attracted unprecedented numbers of Japanese.

(classical Japanese music drama), the simple lines of Japanese painting, and the delicacy of tea ceremonies delighted Japanese readers.[6]

All this publicity and excitement precipitated a boom in the market for books about Einstein and relativity. *Kaizo* (Restructuring) and other magazines published numerous articles on Einstein and his theory of relativity and, for several years, publishers released books on relativity theory "in rapid succession."[7] Jun Ishiwara, who had been Einstein's interpreter during the visit, edited a book containing Einstein's lectures, addresses, and some personal anecdotes (see Figure 3.2).[8] He later edited a Japanese edition of the four volumes of Einstein's completed works.

Einstein's visit and the "relativity boom" had a profound impact on young Japanese physics students. Two future Nobel Laureates in physics, Yukawa and Tomonaga were middle school students in Kyoto when Einstein delivered his public lectures there on the origin of the theory of relativity. In his autobiography, *Tabibito* (*The Traveler*), Yukawa remembered:

> Einstein went from Tokyo to Sendai, then back to Tokyo, and returned to Kyoto in December. He lectured there before a full house, although Kyoto was a place where people seldom gathered in groups. The lecture was on that difficult subject called the Principle of Relativity. Perhaps even the people of Kyoto, who usually showed no quick affection, were attracted by Einstein's personality; or it may be their peculiar quality of seizing upon new things that brought them out this time. No, it cannot be those things. The Theory of Relativity and its originator had been a topic of conversation in all civilized countries of the world, and neither Japan nor Kyoto could be exceptions.[9]

FIGURE 3.2 Jun Ishiwara edited Einstein's lectures for publication in Japan (above). The book was beautifully illustrated by the famous cartoonist, Ippei Okamoto. Below left: Einstein and a deer bowing each other. Below right: Einstein attending the tea ceremony.

Yukawa did not attend that lecture, but many other young science students did. One middle school student in Kyoto who attended Einstein's lecture later recalled that people's "curiosity [was] piqued by all the media coverage. We went without really knowing what to expect."[10] Tomonaga also was swept up by the excitement:

> When I was in the fifth grade of middle school, the famous Prof. Einstein came to Japan. The newspapers devoted much space to him without understanding his theory to any

depth. My imagination was captured by these articles. I delved into a book by Dr. Jun Ishiwara. Being a cocky middle school student, I was fascinated by the relativity of time and space, the four-dimensional world, and non-Euclidian geometry. How mysterious the world of physics was, I thought, and how wonderful it would be to study such a world.[11]

Einstein provided fledging Japanese physicists with a new and radically different image of what a physicist could be. Einstein did not limit his talks to very specific subjects like most Japanese physicists; instead, he talked about general themes, such as space, time, and the universe, weaving together simple language and difficult mathematics. In his lecture of December 14, "How I Created the Theory of Relativity," his particular genius stood out and indicated how different his approach was from that of his Japanese counterparts. Japanese physicists pursued specific data, the results of tedious and time-consuming experiments and calculations, but Einstein insisted that his revolutionary ideas usually came "suddenly":

> I spent almost a year in vain trying to modify the idea of Lorenz in the hope of resolving [the special theory of relativity]. By chance a friend of mine in Bern (Michele Besso) helped me out .... *Suddenly* I understood where the key to this problem lay .... Within five weeks the special theory of relativity was completed .... My first thought on the general theory of relativity was conceived two years later, in 1907. The idea occurred *suddenly* .... The breakthrough came *suddenly* one day. I was sitting on a chair in my patent office in Bern. *Suddenly* a thought struck me ....[12]

Einstein's visit provided the Japanese people with an opportunity to reconsider the role of the physicist. Einstein seemed not merely a physics specialist, but the modern equivalent of the legendary Oriental sage. In 1921, Ishiwara's popular book about Einstein, *The Theory of Relativity*, was published, in which the author discussed "the theory of relativity and philosophical problems," "natural science and philosophy," and "laws in natural science and absolute universality."[13]

Then came the news of revolutionary developments in physics: in 1923, Louis de Broglie's theory on the wave nature of the matter and Compton's experiment of the scattering of x-rays by electrons; from 1925 to 1928, Heisenberg and Pauli's new quantum mechanics using the matrix technique, Schrödinger's equation for wave mechanics, Dirac's general formulation of quantum mechanics and his relativistic equation for the electron. Because the Japanese physics community was a part of the world physics community, and not as isolated from the center as often suggested, it was only a matter of who would implement this revolution in Japan, and when.

In Japan, the initial reaction to this intellectual revolution was surprisingly muted. Some Japanese physicists realized that they were entering an era of uncertainty but they were a minority and their voice failed to move the Japanese physics community. Between 1925 and 1930, major physics journals in Japan published almost no theoretical papers addressing the new quantum mechanics.[14] Nor was any systematic instruction of the subject offered in any university physics department.

Similarly, Riken's physics program was dominated by experimental and applied subjects throughout the 1920s. Thus, the laboratories of K. Honda, M. Masima, and M. Okochi focused on the properties of various metals, such as magnetic and electric properties of nonferrous alloys, the crystallographical structure of metals, and the production of metals. Kujirai–Setoh's laboratory conducted research on insulating materials and also on the electrolytic oxidation of aluminum. T. Suzuki's laboratory developed a process for making metallic aluminum. I. Wada's laboratory determined the various elements contained in iron, steel, and ores, and also separated rare metallic elements.[15]

Only a few physics laboratories in Riken paid some attention to basic and theoretical research during the 1920s. Nagaoka and his students worked on the spectroscopy and Zeeman effects of various elements and on various theories to account for them.[16] Takamine's laboratory "[was] almost entirely devoted to spectroscopic investigations," and S. Nishikawa's laboratory investigated x-rays and cathode rays.[17] One exceptional figure was Seishi Kikuchi who was working on the cutting-edge topic in Nishikawa's laboratory. Kikuchi showed the wave nature of the electron with clear pictures in his 1928 paper, "Diffraction of Cathode Rays by Mica," which was mentioned in Heisenberg's *Die Physikalischen Prizipien der Quantentheorie* (Figure 3.3).[18]

One reason for this apparently tepid response to the physics revolution that took place in the West during the 1920s was Japan's geographical distance from the center of that revolution. Very few Japanese had personally experienced this intellectual upheaval. Even Japanese scientists who studied abroad, including those at Bohr's Copenhagen Institute, generally spent only a few months investigating specific research topics. Almost all Japanese physics professors and students interested in recent developments in physics could learn about them only through Western journals, leaving those researchers at least several months behind their European colleagues.

Recent studies by historians of science, however, have revised the popular conceptions that there was almost no effort to introduce the new quantum mechanics in Japan before 1928 and that Nishina was responsible for the introduction of the new

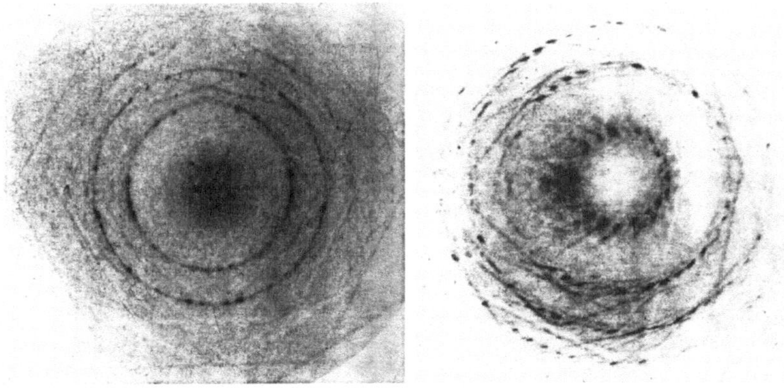

FIGURE 3.3 Two photographs from Seishi Kikuchi's "Diffraction of Cathode Rays by Mica," *Japanese Journal of Physics*, 5 (1928), 83–96 (Figure 3.5 and Figure 3.6). Kikuchi was the leading experimental physicist who excelled other Japanese experimental physicists during the 1920s. From the mid-1930s, he led the physicists at the newly established Osaka Imperial University to build accelerators.

quantum mechanics into Japan.[19] In 1923, Ishiwara introduced the old quantum theory and other contemporary developments to Japanese readers with the publication of his book, *Fundamental Problems in Physics*.[20] From 1924, several small study groups emerged in Tokyo, Kyoto, and Sendai to study the recent development of physics in Europe. The weekly colloquium at the Department of Physics in Tokyo Imperial University was one of such places. Ukichiro Nakaya, a third year physics student in 1924, retained a strong recollection of some of the debates that occurred during the process:

> One evening in June, Nagaoka sensei [teacher] introduced a newly arrived Bohr's book [in fact, Helge Holst and H. A. Kramer's book on Bohr's theory] and explained for about an hour an outline of Bohr's atomic theory, which was quite new at that time. Since we already knew the normal [hydrogen's] orbits and Planck's quantum theory and professors had thought about them for a while, we started a discussion rightly after the presentation. Terada sensei commented that "If an electron goes to the next orbit, it will emits light with frequency ν, and if it jumps again over the next orbit, it will emit light with frequency ν'. If this is true, it means that the electron knows which orbit it should go to and emit light with a specific frequency." For this queer comment, Nagaoka sensei became dumb with smile. Sano Shizuo went to the blackboard and explained "Look. When an electron jumps to another orbit, it hangs around there for a moment, and it emits light during that short period. It hangs around there just for a moment. It's true!" He banged the blackboard with a chalk as if he would like to show how the electron is hanging around for a moment. Terada did not accept the explanation and said, "According to the idea of stationary states, the electron cannot emit light while it is on the orbit. That would contradict the fundamental concepts." Dr. Takahashi Yutaka stood up slowly and said, "If we suppose that an electron departs from an orbit with a specific tangent and reaches another orbit by drawing a spiral, the angle of this tangent determines ν and also the orbit the electron should go." Terada sensei however was obstinate, saying that "I cannot agree to such an arbitrary idea."[21]

At Tokyo Imperial University throughout the 1920s, new subjects in physics were not regularly taught.[22] Nagaoka often taught "quantum and relativity theories" for third year students and lectured on the subjects of previous Solvay Conferences. Kanetaka Ariyama and Toshinosuke Muto remembered that in their third year (1927) there was no separate course on the new quantum mechanics and that it was only briefly mentioned. Students were largely left to study it by themselves in the library. They called this self-study "quantum mechanics game" and most students in theoretical physics participated in it. Only in 1928 did Takujo Itai begin to offer a separate lecture course on quantum mechanics to third year students. It was however not based on deep understanding of the subject nor on the professor's own research. In the Department of Physics at Tokyo Imperial University, not a single faculty member did seriously research in the new quantum mechanics.

The situation in Kyoto was not far different from that in Tokyo. There was no separate course for the new quantum mechanics. As Yukawa remembered:

> In 1926, my first year at the [Kyoto Imperial] University, Erwin Schrödinger introduced wave mechanics and created a sensation in physics .... The physicists were excited and

> Japan, too, began to feel the waves. I guessed what was happening from conversations between the professors and my seniors and I felt that I could not afford to be idle.
>
> In my second year, I spent all my spare time in the physics library. I had no use for the old books that filled the shelves but wanted to learn, as quickly as possible, the articles that concerned the new quantum theory, those that had been published in foreign, especially German, journals within the past two or three years . . . . Soon I decided to read Schrödinger's own papers systematically, as they were the easiest to comprehend.[23]

Around 1927, a few students at Kyoto Imperial University, including Yukawa and Tomonaga, formed an informal study group. Tomonaga remembered:

> There were, however, some ambitious senior students studying quantum mechanics on their own. They let me join and gave various advices. Mr. Shohei Tamura and late Mr. Sotohiko Nishida were the leaders of these ambitious modern boys.[24]

They read Sommefeld's *Atombau und Spektrallnien* and articles in the *Annalen der Physik* and in the *Handbuch der Physik* in the library without any help from faculty members. A young professor of theoretical physics, Kajuro Tamaki, who had worked in relativity theory and fluid mechanics, provided these students with protection and encouragement. Another professor, Masamichi Kimura, arranged for Takeo Hori of Tohoku Imperial University to deliver some lectures at Kyoto on molecular spectroscopy.[25] Later he invited Sugiura (1930) and Nishina (1931) to Kyoto to give a special series of lectures on the new quantum mechanics.

Around 1928, a small physics group at Tohoku Imperial University in Sendai began to study "Theories of Solid and Electron [*sic*]."[26] This group included Professors Yutaka Takahashi and Hiro Yamada along with the young physicists, Tadayoshi Hikosaka, Takeshi Hayasi, and Rikuo Nakabayashi. Tokutaro Hirone, who graduated from Tohoku Imperial University in 1928, remembered that he had attended the lecture course, "Quantum Theory," given by Takahashi.

However, in terms of the future development of quantum mechanics in Japan, the most important informal study group was organized in 1926 at Riken by six young physicists in the Tokyo area who were disappointed by their senior professors' lack of enthusiasm for quantum mechanics.[27] Twelve physicists participated in the first meeting, which was held on the evening of March 18. Of these, only one, Mitsuharu Fukuda, was a graduate of Kyoto Imperial University. The rest of the newly formed "Physics Reading Group," as it became known, were graduates of Tokyo Imperial University who held positions at Tokyo University or at Riken and were interested in atomic physics or related subjects. The six leading members who organized the first meeting were Uzumi Doi, Masamichi Konko, Kamekichi Shiba, Akira Suzuki, Ukichiro Nagaya, and Yoshio Fujioka, all fairly recent graduates of the Department of Physics in Tokyo Imperial University. Three members, Torahiko Terada, Shoji Nishikawa, and Masazou Kiuchi were both professors of experimental physics at Tokyo Imperial University and chief researchers at Riken. Two members, Jirô Sasaki and Isamu Nitta, were graduates of the Department of Chemistry and were working at Riken when they joined the group. Soon Seishi Kikuchi, Chuji Tsuboi, Yutaka Takahashi, Kotaro Tomiyama, Taro Suga, and Yasumasa Tani also joined.[28]

The Physics Reading Group read, discussed, and critiqued recent papers about atomic physics published in books or in such journals as the *Zeitschrift für Physik* and the *Annalen der Physik*. The central focus of their reading and of their lively, often heated, discussions was the new quantum mechanics.[29] Terada was so impressed by the young physicists' passion to learn about new developments in physics that he criticized the traditional Japanese university education:

> The university is not the place to put knowledge into students' heads but to teach them the method of study or to make them interested in study itself. The true study starts only after graduation. If we teach them how to walk, the students will walk away wherever they want to go after graduation. It is bad to load students with unreasonably heavy burden without teaching them how to walk.[30]

For the young "student radicals" who formed the Physics Reading Group, the group's success was a truly impressive triumph over their senior physics professors in the imperial universities.[31] After a year of intense reading and weekly discussions, the group's members collected some important papers on the new quantum mechanics and related subjects, summarized them when necessary, translated them into Japanese, and published them, in 1927, as *Readings in Physics*.[32] This was the first Japanese-language book on quantum mechanics that included critical works by many founders of the new quantum mechanics, including de Broglie, Heisenberg, Schrödinger, Dirac, Born, Jordan, and others. The Group published a second volume of *Readings in Physics* the following year.

However, after 1928, the Physics Reading Group lost momentum. Without a visionary senior leader, the volunteer members of the group were no longer making much headway. By the time Nishina returned from Europe and became acquainted with its members, the group's size had dwindled and its liveliness had subsided. By the early 1930s, the group's activities had devolved into lunchtime discussions of nuclear physics and cosmic rays, organized by the Takamine laboratory.[33]

In 1927, the Physico-Mathematical Society of Japan founded a new journal in Japanese to introduce "important papers published in Japan and abroad" and to provide "better communication to its members."[34] More than one third of the 26 papers that appeared in the first volume concerned the new quantum mechanics: for example, Heisenberg's "*Über quantentheoretische Umdeutung kinematischer und mechanischer Beziehungen*," Born and Jordan's "*Zur Quantenmechanik I*," Born, Heisenberg and Jordan's "*Zur Quantenmechanik II*," Fritz London's "*Die Theorie von Weyl und die Quantenmechanik*," and Bohr, Kramer, and Slater's "*The Quantum Theory of Radiation*."[35]

In December of 1928, Sommerfeld visited Japan. This was perhaps the last event before Nishina's return to Japan that was significantly important to Japan's introduction to the new quantum mechanics. Sommerfeld delivered several lectures on wave mechanics and electron theory at Riken and at Tokyo Imperial University on December 5, 6, 7, and 19. In attendance at these lectures was Ariyama, who was deeply affected by Sommerfeld's presentations:

> After attending Sommerfeld's lectures, I became familiar with quantum mechanics for the first time. Until then, quantum mechanics had meant to me something related with atoms or molecules or something abstract only related with a big issue like the properties

> of matter. But after his lectures it suddenly became a familiar thing directly related with the matter around me. For the first time I could connect Lorentz's electron theory that I had learned in high school with Sommerfeld's theory. I felt strong affinities toward quantum mechanics.[36]

Thus, it is clear that interest in the new quantum mechanics had been sprouting in Japan for quite some time before Nishina's return from Europe in December 1928. A number of younger Japanese physicists had realized that the future would belong to this new field and were eager to study it. Clearly, as part of the international physics community, the Japanese physics community would inevitably develop an interest in the new quantum mechanics, but the country lacked serious researchers and experienced teachers in this field. Yoshio Nishina, a man who had been at the center of this intellectual revolution in Europe, at this very moment, on his return to Japan, was bringing with him the experience and skills needed to become the leader of the "quantum generation" of Japanese physics.[37]

3.2 NISHINA: THE PREACHER OF THE NEW QUANTUM MECHANICS

For Nishina, who arrived in his homeland on December 21, 1928, the next few years were a mixture of disappointment and hope. Despite his brilliant work in Europe, which certainly surpassed that of any other Japanese physicists of his generation, not a single Japanese university offered him a professorship. Japanese tradition, which dictated that a scholar's origin determined the rest of his academic career, stood in the way. An academic position, whether that of *koza* (the Japanese professorship roughly equivalent to the European chaired professorship), associate professor, or assistant, was filled by a graduate of the department having the vacancy, and promotion was automatic except in the case of entirely new posts. Within this rigid system, Nishina had little hope of being offered a physics professorship. As a graduate of Tokyo Imperial University, he had little hope for a position at another university: the old days when the vacant or new posts at other universities were mostly filled by the graduates of Tokyo Imperial University had gone forever. Moreover, as a graduate of Tokyo Imperial University's Department of Electrical Engineering, he would not be offered a professorship by that university's Department of Physics. Nishina was too old, at 39 in 1929, to be appointed to a junior-level post. His only hope for an academic career in Japan was to be offered a newly instituted professorship, perhaps in quantum theory, at one of Japan's imperial universities or at a brand-new university with no graduate pool from which to hire. However, because quantum theory was not yet recognized as an independent field of physics in Japan, this possibility was remote. Even the very influential Nagaoka, who had been a most generous patron of Nishina, could not change the situation. All Nagaoka could do was relocate Nishina from Kujirai's laboratory to a position as researcher at his own laboratory in Riken. Masa Takeuchi, one of Nishina's first assistants, remembered that Nishina frequently played tennis with Sanae Yoshida from Nishikawa's laboratory in those days.[38]

Despite his long stay in Europe, Nishina obediently followed Japanese customs when he returned home. For the first few months after his long absence, family matters demanded most of his attention and energy. In a letter to Hevesy on April 1, 1929, he wrote: "I have had to deal with a good deal of family affairs which have accumulated during [the] last eight years and the duty of which I have neglected to do. There is an endless chain to follow."[39] The most important of these duties was his obligation to marry. He married Miss Mie Nawa, who was a sister of Nishina's close friend, Takeshi Nawa, on February 23, 1929, just 2 months after he returned home.[40] He wrote to Hevesy:

> [The] last thing which has taken my time has been a marriage question and its consequences. *I have had to marry about 5 weeks ago.* You will perhaps be surprised that such an important affair of life has been carried out in such a short time. It is nearly impossible from [the] Western view-point, I myself did not know at the time of my arrival that I should marry in such a short time. But each country and even each family has its own custom and tradition and I am after all a man who was born in Japan, and according to the advice of my family and friends, I decided to marry a sister of my intimate friend here and had the wedding on the 23rd Feb. At the same time I rented a small house in order to form my home. I have had to furnish and arrange this house. I have had the first experience of running a house. All these have taken whole my time. You will see why I have not written a line to you so long.[41]

Hevesy immediately wrote back with warm congratulations and advice:

> Married life means also a great change in one's life. You have to adapt your ways and moods to family requirements. However it makes one more settled and brings one nearer to other human beings. You understand easier their wants, wishes, pleasures and sorrows.[42]

Shortly after his wedding, Nishina wrote Bohr:

> I am having a quite satisfactory life after marriage, but the marriage has had its consequences which have taken whole my time since then: All sorts of formalities, arrangements, family affairs and so on. I have thus been unable to do any physics at all, but hope soon to get into a serious work again.[43]

Yoshio and Mie produced two sons — Yuichiro and Kojiro.

In the face of his family obligations, Nishina's research at Riken was not as impressive as it had been in Copenhagen. He completed two papers on the polarization of Compton scattering that he had begun there and submitted them to *Nature* and to *Zeitschrift für Physik*. Although his official job in Nagaoka's laboratory was "quantum theory and its applications," he wrote only a few papers on the subject, including a Japanese-language introduction to the problem of causality in quantum mechanics for electrical engineers.[44] Instead, he returned to an old subject, x-ray spectroscopy, and read some papers, in Japanese, on that subject at Riken's semiannual meetings.[45] He also wrote a book on x-ray spectroscopy for Japanese experts.[46] After completing a thesis that modified his 1925 paper, "On the L-Absorption Spectra of the Elements from Sn (50) to W (74) and their Relation to the Atomic Constitution," he received a doctoral degree of science from Tokyo Imperial University on November 21, 1930.

Compared to his research at this time, Nishina's contributions to the Japanese physics community in terms of "preaching [the] new quantum mechanics" and bridging the gap "between Japanese physicists and important foreign dignities" were quite impressive.[47] He played a central role in bringing important Western physicists to Japan, and delivered lectures on the new quantum mechanics to the next generation of Japanese physicists at several universities across the country.

While still in Europe, Nishina had planned to invite several famous Western physicists to Japan. His first target was his mentor, Niels Bohr. During a skiing trip in January of 1928, Nishina heard that Bohr wanted to travel to Asia.[48] Nishina immediately wrote to Nagaoka that he would like to arrange a visit to Japan for Bohr. In the same letter, he mentioned the possibility of also hosting Heisenberg, Dirac, and Pauli. Nishina's 1929 letters to Bohr indicate that Nagaoka and Nishina had every intention of inviting Bohr to Japan.[49] However, Bohr's visit was postponed for almost a decade. Nishina's Copenhagen connection did enable him to arrange for the 1929 visits of Heisenberg and Dirac. Hevesy came in 1931 and Bohr finally arrived in 1937.

Nishina's enthusiasm for bringing foreign physicists to Japan grew from his desire that Japanese physicists personally meet the scientists who were creating the new fields in physics, especially quantum mechanics. When, in April of 1929, he wrote to invite Dirac to visit Japan, Nishina expressed this wish:

> We hope, that you will be able to give us here four or five lectures and *some opportunities of having a close contact with you and with [the] new physics*. We hope too that you will stay in Japan as long as possible.[50]

FIGURE 3.4 When Werner Heisenberg (fourth from the left) and P. A. M. Dirac (third from the right) visited Japan in the fall of 1929, their lectures on new quantum mechanics stimulated the minds of young Japanese physicists. Nishina (first from the left) carefully arranged their visit, translated their lectures for Japanese audiences, and later edited a book containing these translations. (Courtesy of the Institute of Physical and Chemical Research.)

After Heisenberg and Dirac's departure from Japan, Nishina reported to Bohr that they had given "several lectures each, which were of great benefit to us. To our regret, however, they stayed only 2 or 3 weeks, which were much too short."[51] Nishina reiterated this point in his preface to *Heisenberg and Dirac's Lectures on the Problems of Quantum Mechanics*, which he carefully edited and translated:

> The editor [Nishina] often heard that Japanese scientists in some fields were just imitating Western scientists in the same fields. I agree that the geographical remoteness from the center may be a reason why our science community has been behind the progress. So, it is very important for our science community to invite the scholars who are leading the progress and to [hear] from them the present state of the field. Those who know the rapid development of physics in America during the recent years will agree with me that this idea is not far from the truth. In this respect, the invitation of two scholars [Heisenberg and Dirac] by the Keimei Kai Foundation is very timely and meaningful one.[52]

Nishina later wrote several introductory articles for the popular science journal, *Kagaku*, when Irvin Langmuir (1934), Dirac and G. Beck (1935), and Bohr (1937) visited Japan.[53]

Heisenberg and Dirac's visit in September of 1929 would be the first major opportunity for Japanese physicists to meet highly successful European physicists who directly contributed to the birth of the new quantum mechanics, and Nishina made every effort to make their visit successful. He knew through his Copenhagen connections that Heisenberg and Dirac would give lectures at American universities in the spring of 1929, and asked them whether they could visit Japan to give some lectures before returning to Europe:

> We should be very much pleased if you would come over here on your way home, so that we all can see you and hear from you some [of the] latest news in physics. Certainly you will not be interested in Japanese physics which is not very strong but our country is unique in some respects and we hope that may interest you.[54]

Nishina persuaded Nagaoka to secure suitable sponsors for the event; the Keimeikai Foundation and Riken agreed to share the budget.[55] Heisenberg and Dirac both accepted Nishina's invitation and, in August, they sailed together from San Francisco to Honolulu to Yokohama.

From September 2 to 7, Heisenberg and Dirac each gave six lectures at Riken and Tokyo Imperial University. They spoke about recent developments in physics, including the fundamental principles of the new quantum mechanics, quantum statistical mechanics, the relativistic theory of the electron, the theory of ferromagnetism, the principles of superposition and the quantum mechanical properties of two-dimensional harmonic oscillators. Details of these lectures are provided in Table 3.1. Needless to say, these lectures profoundly influenced many young Japanese physicists. Tomonaga traveled from Kyoto to Tokyo to attend the lectures:

> In the year 1929, when we were graduated from the university, Heisenberg and Dirac came to Japan in September. They gave lectures in Tokyo and Kyoto. I got up my nerve, went to Tokyo, and listened to the lectures . . . .
>
> Miraculously, I remember, I could more or less understand the content of the lectures because fortunately I had already looked through papers related to these talks. This was

TABLE 3.1
Lectures Given by Heisenberg and Dirac (September 2–7, 1929)

	Heisenberg's lectures	Dirac's lectures
September 2: Riken	The Indeterminacy-Relations and the Physical Principles of Quantum Theory	The Principles of Superposition and the Two-Dimensional Harmonic Oscillator
September 3: Tokyo Imperial University	Theory of Ferromagnetism	The Basis of Statistical Quantum Mechanics
September 4: Tokyo Imperial University	Theory of Ferromagnetism, Continued	Quantum Mechanics of Many-Electrons System
September 5: Tokyo Imperial University	Theory of Conduction (by F. Bloch in Leipzig)	Quantum Mechanics of Many-Electrons System, Continued
September 6: Tokyo Imperial University	Retard Potential in the Quantum Theory (joint work with Pauli)	Relativity Theory of the Electron
September 7: Tokyo Imperial University	Retard Potential in the Quantum Theory (joint work with Pauli), Continued	Relativity Theory of the Electron, Continued

Source: Y. Nishina, *Bulletin IPCR*, *8* (1929), 859–867.

the first time I had come from rural Kyoto to Tokyo and seen in person distinguished people like Professor Hantaro Nagaoka, Professor [*sic*] Nishina, and Professor Sugiura and also brilliant graduates of the University of Tokyo, who obviously looked very bright. I listened to the lectures, hiding myself toward the last row of the room, overwhelmed by those luminaries. There was one senior student who had finished at the Third High School of Kyoto and was graduated from the physics department of the University of Tokyo, and he pointed out to me that that was Professor Nishina and those were Masao Kotani and Tetsuro Inui, who were studying quantum mechanics in Professor Nishina's colloquium. He encouraged me to get acquainted with these people, but I simply stayed shy. In this atmosphere, I clearly remember the question Dirac asked Heisenberg after the third lecture. As you probably know, Heisenberg and Pauli in their theory introduce the condition $\nabla \cdot E = 4\pi\rho$, not as a relation between the q-numbers, but as an additional condition for the state vector ψ, $(\nabla \cdot E)\psi = 4\pi\rho \cdot \psi$. Dirac asked the question whether the eigenvalue 0 of $\nabla \cdot E - 4\pi\rho$ is discrete or continuous. Apparently this was an unexpected question for Heisenberg. He could not give an answer right away and after thinking for a while answered, "It is probably continuous." I remembered on the last day of the lectures at the University of Tokyo, Professor Nagaoka got up and raved about how Heisenberg and Dirac in their twenties had accomplished such a major thing as the establishment of a new theory and deplored that in Japan the physicists were still picking up the chaff and bran of Europe and America, and that students were just copying lectures, which was terrible. "You guys should emulate Heisenberg and Dirac." (Professor Nagaoka ranted this in his Nagaoka-English, and I could not quite hear it accurately, so this may be my arbitrary translation.)[56]

During the six days of lectures, Nishina stood beside Heisenberg or Dirac, translating their lectures into Japanese and adding explanations when necessary. This visible role made it obvious to Japanese physicists that Nishina really knew the new quantum mechanics better than any other Japanese scientists. The fact that Heisenberg and Dirac, two world-famous physicists, treated Nishina as their equal greatly helped overcome Nishina's greatest vulnerability in the traditional Japanese academic system — the fact that he was not a physics graduate.

Because only a very small body of Japanese-language literature dealt with the new quantum mechanics, Nishina decided to take the task of editing and translating the lectures of Heisenberg and Dirac, and obtained advance estimates from each university of the number of translations each would order.[57] In several footnotes, Nishina added the latest related papers or books on related subjects. For example, to further explore the contents of Heisenberg's first lecture, "The Indeterminacy-Relations and the Physical Principles of Quantum Theory," Nishina recommended reading Heisenberg's 1930 book, *Die Physikalischen Prinzipien der Quantentheorie* (Figure 3.5). The translation finally was published in 1932, and more than 150 copies were distributed to various Japanese universities.[58]

As Nishina worked on his translation of the lectures of Heisenberg and Dirac, he arranged a visit to Japan by George de Hevesy, who arrived in the spring of 1931 (Figure 3.6). Hevesy, who was an authority on radioactivity and the pioneer of the radioactive tracer, had informed Nishina that, with his wife, he would be

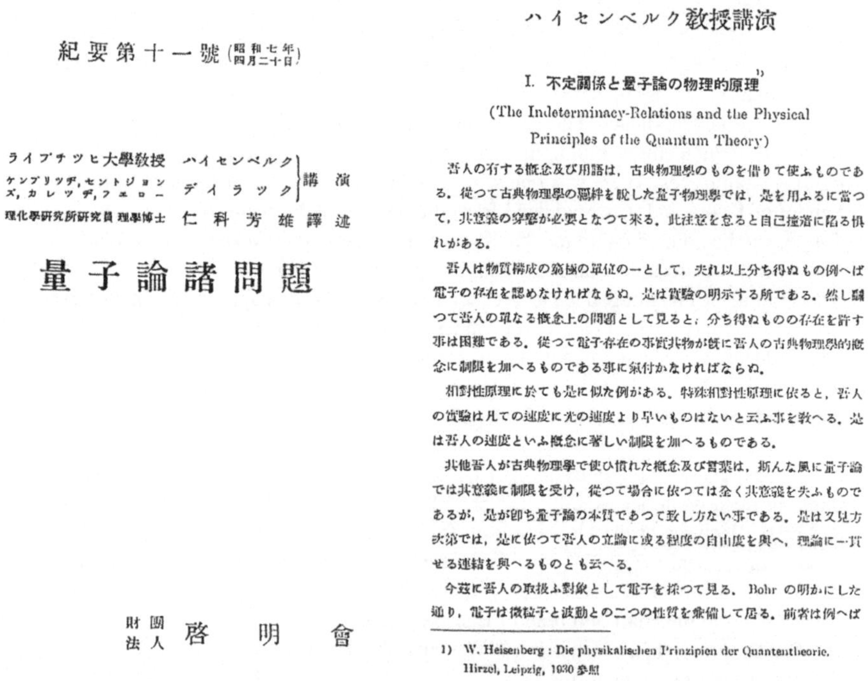

紀要第十一號（昭和七年四月二十日）

ライプチッヒ大學教授 ハイセンベルク
ケンブリッヂ，セントジョンズ，カレッヂ，フェロー ディラック
講演
理化學研究所研究員 理學博士 仁科芳雄譯述

量子論諸問題

財團法人 啓明會

ハイセンベルク教授講演

I. 不定關係と量子論の物理的原理[1)]

(The Indeterminacy-Relations and the Physical Principles of the Quantum Theory)

吾人の有する概念及び用語は，古典物理學のものを借りて使ふものである。從つて古典物理學の羈絆を脱した量子物理學では，是を用ふるに當つて，其意義の穿鑿が必要となつて來る。此注意を怠ると自己撞着に陥る惧れがある。

吾人は物質構成の窮極の單位の一として，夫れ以上分ち得ぬもの例へば電子の存在を認めなければならぬ。是は實驗の明示する所である。然し翻つて吾人の單なる概念上の問題として見ると，分ち得ぬものの存在を許す事は困難である。從つて電子存在の事實其物が既に吾人の古典物理學的概念に制限を加へるものである事に氣付かなければならぬ。

相對性原理に於ても是に似た例がある。特殊相對性原理に依ると，吾人の實驗は凡ての速度に光の速度より早いものはないと云ふ事を教へる。是は吾人の速度といふ概念に著しい制限を加へるものである。

其他吾人が古典物理學で使ひ慣れた概念及び言葉は，斯んな風に量子論では其意義に制限を受け，從つて場合に依つては全く其意義を失ふものであるが，是が卽ち量子論の本質であつて致し方ない事である。是は又見方次第では，是に依つて吾人の立論に或る程度の自由度を與へ，理論に一貫せる連結を與へるものとも云へる。

今茲に吾人の取扱ふ對象として電子を採つて見る。Bohr の明かにした通り，電子は微粒子と波動との二つの性質を兼備して居る。前者は例へば

1) W. Heisenberg : Die physikalischen Prinzipien der Quantentheorie. Hirzel, Leipzig, 1930 參照

FIGURE 3.5 "Heisenberg and Dirac's Lectures on the Problems of Quantum Mechanics," *Keimeikai Gaiyo*, *11* (1932), 3, which Nishina edited and translated into Japanese, was published in 1932.

residing in the United States in 1930–1931 and could visit Japan on the way home.[59] Nishina secured 3000 yen to fund a "few lectures" by Hevesy to Japanese scientists.[60] From March 23 to March 25, Hevesy gave the following lectures at Riken: "The Discovery and Properties of Hafnium," "Transport of Matter through Crystalline Bodies," and "Radioactive Indicators in Chemistry, Physics and Biology." From March 26 to March 28, at the Department of Chemistry at Tokyo Imperial University, his lectures were "Quantitative Analysis by X-Rays," "Separation of Isotopes and the Radioactivity of Potassium," and "The Abundance of the Elements." On April 1 at the Department of Chemistry of Kyoto Imperial University, he lectured on "Quantitative Analysis by X-Rays." The next day, he gave another lecture there entitled, "The Age of Earth."[61] Some of these topics later became important research subjects at Nishina's own laboratory at Riken.

In April 1937, Bohr finally visited Japan with his wife, Margrethe, and son, Hans (Figure 3.7). Nishina had prepared for this visit since 1929, but it had been delayed because of Bohr's busy schedule and by the sudden death of his eldest son, Christian, in the summer of 1934.[62] The Keimeikai Foundation, which had supported Heisenberg–Dirac's visit in 1929, once again provided financial support for the visit. During his month-long stay in Japan, Bohr gave lectures in quantum mechanics, nuclear physics and even some philosophy. He chose "Principles of Atomic Theory"

FIGURE 3.6 George de Hevesy (the fifth from the left in the front row) and Mrs. Hevesy (the seventh from the left in the front row) visited Riken. Nishina (the first from the right in the second row) and Mrs. Nishina (the fourth from the right in the first row) as well as the president of Riken, Okochi (the sixth from the left in the front row), Nagaoka (the third from the right in the front row), and Suguira (the first from the left in the third row) greeted the couple. (Courtesy of the Institute of Physical and Chemical Research.)

FIGURE 3.7 Niels Bohr's visit to Japan in the spring of 1937. It reconfirmed Nishina's leading role in the Japanese physics community. In the right photograph, Bohr (right) was discussing something with Nishina (left) and Seishi Kikuchi (middle). (Courtesy of Niels Bohr Archive [left] and AIP Emilo Segrè Visual Archives [right].)

TABLE 3.2
Niels Bohr's Lectures ("Principles of Atomic Theory") at Tokyo Imperial University (April 19–28, 1937)

April 19	General Lecture: Development of the Theory of Electron (Historical Review)
April 20	General Lecture: Problems in Nucleus
April 21	Professional Lecture: Fundamentals of Quantum Mechanics
April 22	Professional Lecture: Mathematical Expression of Quantum Mechanics
April 24	Professional Lecture: Problems in Scattering
April 27	Professional Lecture: Problems in Nucleus
April 28	General Lecture: Quantum Mechanics and Its Relation to Philosophy; its applications to Biology and Psychology

Source: The History of Physics in Japan, Vol. 1, p. 306.

as the subject for the series of lectures at Tokyo Imperial University (April 19–28), "The Structure of Nucleus" for Tohoku Imperial University (May 3), "Atomic Nuclei" for Kyoto Imperial University (May 10), and "Causality in Atomic Theory" for Osaka Imperial University (May 12).[63] Table 3.2 summarizes the lectures given at Tokyo Imperial University. Nishina accompanied Bohr and stood beside him, translating his lectures into Japanese and adding his own explanations when necessary. He also wrote some articles about Bohr and his lectures for general readers.[64] Bohr's visit reconfirmed Nishina's prominence in the Japanese physics community.

Bohr's 1937 visit to Japan was different from Heisenberg–Dirac's 1929 visit in several ways. First of all, Bohr's visit made little impact on the Japanese physics community. As Chapters 4 to 6 will demonstrate, the Japanese physics community had grown so rapidly since the beginning of the 1930s that Japanese physicists could not only readily understand Bohr's lectures but also offer their own ideas and opinions to the father of quantum mechanics. In other words, there was not another Tomonaga who had been deeply influenced by and remembered Heisenberg–Dirac's lectures. Second, although most of Bohr's lectures dealt with scientific matters, some discussed the philosophical meanings of quantum mechanics and the applications of quantum mechanics to other branches of science. Ito indicates that Bohr's lectures aroused hot debates among Japanese philosophers in the late 1930s.[65]

While playing a central role in bringing important physicists to Japan, Nishina started delivering his own lectures on quantum mechanics. A few months after Nishina's return from Europe, he wrote to Hevesy:

> [The] next thing on which I have had to spend a good amount of my time is the lectures I have had to give on current problems in physics. To prepare such lectures is one thing in itself, but to write them down in a complete form for publication is another thing which is not so easy.[66]

The first institution to invite Nishina to lecture was Tokyo Bunrika University, where he lectured part time on the new quantum mechanics from April of 1930 to March of 1933. In 1931, Masamichi Kimura invited him to give a special series of

FIGURE 3.8 Bohr and Mrs. Bohr visiting Nishina's home. (Courtesy of Niels Bohr Archive.)

lectures on the new quantum mechanics at Kyoto Imperial University, and shortly after, so did Hokkaido Imperial University. It was not until 1937, however, that the Department of Physics at Tokyo Imperial University finally relented and asked Nishina to become a *part-time* lecturer there.[67]

Nishina's month-long lecture series on new quantum mechanics at Kyoto Imperial University was especially important to the development of physics in Japan. The future leaders of the Japanese physics community, namely, Yukawa, Tomonaga, Shoichi Sakata, and Minoru Kobayasi, all attended Nishina's lectures, met him personally, and were strongly influenced by him (see Figure 3.9).[68] Kimura, who organized the lecture series, was a specialist in spectroscopy whose own European sojourn had made him aware that Japan needed to learn about the new quantum mechanics as soon as possible. Previously, in May of 1930, at the request of Kimura, Sugiura had presented a series of lectures on the applications of quantum mechanics, which he had studied in Copenhagen and Göttingen. During his European stay, Sugiura had published three papers: (1) about his application of the principles of quantum mechanics to helium and lithium atoms; (2) about his application of Heitler–London's method to the hydrogen molecule; and (3) demonstrating his calculation of the rates for electronic transitions in sodium by applying Schrödinger's wave mechanics.[69] He returned to Japan in 1927 and resumed work in Nagaoka's laboratory

FIGURE 3.9 Nishina (the second from the right in the first row) gave a month-long series of lectures on the new quantum mechanics at Kyoto Imperial University. In attendance were two future Nobel Laureates in physics, Hideki Yukawa (the second from the left in the standing row) and Sin-Itiro Tomonaga (the third from the left in the standing row), who were greatly influenced by Nishina. (Courtesy of the Institute of Physical and Chemical Research.)

at Riken, where, unfortunately, he never again attained the high productivity he had in Europe and soon disappeared "from Japan's scientific scene."[70] Although Sugiura tried to help Yukawa and Tomonaga by proposing research projects, his suggestions did not prove helpful. In his memoir, Tomonaga recounted:

> The project [Professor Sugiura] gave me was the problem of the Na_2 molecule (I do not know whether he knew that I was interested in molecule [*sic*]), which was the task of applying Heitler and London's theory of H_2 to Na_2. I was already lukewarm about molecules at that time, but I thought it would be instructive to do something by myself rather than reading other people's papers, and therefore, I dared to say, "Let me do it." This was a numerical calculation from the beginning, and it did not look that instructive. (It was good training for perseverance, however.) Moreover, it was like picking up the crumbs of Professor Sugiura's work, and it was not inspiring at all. Also, when the calculations were carried out, there appeared many awkward results, and the whole thing did not work.[71]

Yukawa, whom Sugiura advised to theoretically "explain the peculiar result of [the] Bergen Davis' experiment," met similar difficulties. Both future Nobel Laureates felt "stuck" for a year in the situations recommended to them by Sugiura.[72]

Nishina's 1931 series of lectures at Kyoto Imperial University had quite different results than those of Sugiura's lectures the previous year. Nishina employed the text of Heisenberg's *Die Physikalischen Prizipien der Quantentheorie* (1930), and he emphasized the fundamentals of the new quantum mechanics. In his memoir, Tomonaga neatly summarized the impression made by Nishina on the young Kyoto physicists:

> I remember that Nishina sensei's lecture lasted about a month. But, during that short time, he left very strong impression to us. The lecture consisted of both physical and philosophical contents which made the entangled ideas very clear. Particularly, I can't forget the discussion after the lecture.
>
> We had already studied the Klein–Nishina formula. Young students wondered what kind of man is he who put his name to a formula. But, on the other hand, many such world-class scientists gave pressure to young students. It was never easy for young students, who were often obsessed by inferior [*sic*] complex, to ask a question or to explain their ideas to such great teachers.
>
> However, Nishina looked far from a world-class scientist we imagined (we guessed a sharp looking figure). He looked mild and talked without any sharp remarks. So, after many hesitations and thoughts, we asked questions and tried to explain our ideas to him.[73]

Nishina's treatment of students was very different from that of most other Japanese sensei (teachers). His gentle approach was somewhat shocking to young Japanese physicists, and they were emotionally touched by his consideration for their feelings and work. Tomonaga, despite his timid reluctance to talk "on such an occasion," was reassured enough by Nishina's mild manner to dare to ask a question, a circumstance that Western readers unfamiliar with the traditional arrogance of Japanese university professors may have trouble understanding.

Nishina did not disappoint Tomonaga. He "was extremely kind in answering [Tomonaga's question] and listening to" the novice physicist.[74] Nishina even invited Tomonaga and Yukawa to join him for further conversation at dinner. There, the two young physicists explained the year-long difficulties they had experienced attempting to follow Sugiura's recommendations for their research projects.[75] Nishina responded with sound advice, not orders. He informed Yukawa that Bergen Davis' results had been proven wrong and suggested that Yukawa stop his research right away. Nishina showed Tomonaga a letter he recently had received from Bohr and reminded Tomonaga that "many other interesting things" were available to investigate. Nishina's considerate behavior meant so much to Tomonaga that he later wrote a special article for the memory of Nishina with the title of "Nishina Sensei's Warm Heart Moved Me."[76]

To Nishina, the two years following his return to Japan were not only difficult but also very important. Despite his brilliant work in Europe and Bohr's strong recommendation, he was prevented from obtaining a professorship in physics by the rigid

Japanese university system. Even newly opened Osaka Imperial University, whose first president was none other than Nagaoka, did not provide Nishina with a post, for reasons that are unknown even today. Meanwhile, much of Nishina's time was being consumed by his family's need that he had neglected over the past decades. As he frequently complained to his Western friends, he was "unable to do any physics at all" and was "kept away by various external circumstances from getting into real physics."[77] To make matters worse, the research papers he did manage to write during this time were somewhat disappointing.

On the other hand, Nishina had begun to shape his future role in the Japanese physics community and emerged as his homeland's "preacher" of the new quantum mechanics. He was well on his way to becoming Japan's most respected and influential teacher, and he had made a great contribution to the future development of physics in Japan by introducing Heisenberg, Dirac, and Hevesy to his fellow Japanese physicists. Realizing that no Japanese university was likely to violate tradition by awarding Nishina a professorship, some senior scientists who understood that Nishina was too valuable to be kept in a minor position in the Japanese scientific community sought other suitable positions for him. Finally, Okochi, Riken's enlightened director, with the support of many prominent senior scientists, promoted Nishina to a position of chief researcher at Riken. On July 1, 1931, Nishina opened his own laboratory at Riken, the Institute's first chief researcher who was not also a university professor.

Japanese scholars and Nishina's former students often have suggested that he did not want to be burdened by faculty responsibilities and would not have welcomed a teaching post at an imperial university. However, *Nishina enjoyed teaching*, as evidenced by his decision to continue teaching even after he opened his own laboratory at Riken. From 1931 to 1933, he delivered highly influential lecture courses on quantum mechanics at Tohoku Imperial University.[78] He also lectured at Tokyo Bunrika University until 1933 and at Tokyo Imperial University from 1937 to 1942. His devotion to teaching was perhaps best illustrated by the fact that he made his laboratory a forum for young physicists from all over Japan. Without a doubt, Nishina would have been a most impressive professor of physics and an energetic chief researcher at the same time, just as his mentor, Niels Bohr, had been in Copenhagen.

NOTES

1 Atsushi Katsuki, "Before the Returning-Home of Nishina [in Japanese]," *Butsuri*, *45* (1990), 752–754 on 752.

2 Bartholomew, *The Formation of Science in Japan*, p. 254. For more information about the National Research Council, see ibid., Chapter 8, "The Research System in an Age of Transition."

3 The visit was sponsored by Kaizosha, a progressive publishing company. For details of Einstein's visit to Japan, see Stefan Iglhaut (ed.), *Einstein in Japan: A Travelogue* (Tokyo: Dai Nippon Printing Co., 2005).

4 Ayao Kuwaki, *A Biography of Einstein* [in Japanese] (Tokyo: Kaizosha, 1934), pp. 106–116.

5 Jun Ishiwara, *The Record of Einstein's Lectures* [in Japanese] (Tokyo: Kaizosha, 1923; reprint by Tokyo Tosho, 1971), pp. 20–33.

6 Tsutomu Kaneko, "Einstein's View of Japan's Culture," *Historia Scientiarum*, *27* (1984), 51–76.

7 Sigeko Nisio, "The Transmission of Einstein's Work to Japan," *Japanese Studies in the History of Science*, *18* (1979), 1–8 on 6–7.

8 Jun Ishiwara, *The Record of Einstein's Lectures*. A famous cartoonist, Ippei Okamoto, who accompanied Einstein, added many beautiful illustrations to the book.

9 Hideki Yukawa, *Tabibito (The Traveler)*, translated by L. M. Brown and R. Yoshida (Singapore: World Scientific Publishing, 1982), p. 114.

10 Nagamori Ikeda, "Dr. Tomonaga in Middle School," in M. Matsui and H. Ezawa (eds.), *Sin-itiro Tomonaga*, (Tokyo: MYU, 1995), pp. 47–54 on p. 52.

11 Makinosuke Matsui, "Introduction (to Chapter 2. Setting His Heart on Physics)," in Matsui and Ezawa (eds.), *Sin-itiro Tomonaga*, (Tokyo: MYU, 1995), pp. 77–78; S. Tomonaga, "My Teachers and My Friends [in Japanese]," in *Collected Essays of Sin-itiro Tomonaga*, Vol. 1 (Tokyo: Misuzu Shobo, 1981), pp. 193–203.

12 Albert Einstein, "How I Created the Theory of Relativity," translated by Yoshimasa A. Ono, *Physics Today* (August, 1982), 45–47 on 46–47. Italics are added.

13 Jun Ishiwara, *Theory of Relativity* [in Japanese] (Tokyo: Iwanami Shoten, 1921), Introduction.

14 D. Kim, "The Emergence of Theoretical Physics in Japan," *Annals of Science*, **52**, (1995), 394–395.

15 "Summary of the Past Activities of the Institute," *SP*, *34* (1938), 1763–1892.

16 Ibid., 1826–1833.

17 Ibid., 1874–1878 and 1837–1841.

18 S. Kikuchi, "Diffraction of Cathode Rays by Mica," *Japanese Journal of Physics*, *5* (1928), 83–96. Heisenberg mentioned Kikuchi's experiment when he explained the diffraction of electron. See W. Heisenberg, *Die Physikalischen Prizipien der Quantentheorie* (1930), p. 7.

19 Dong Hoon Oh, *Nishina Yoshio and the Modern Physics in Japan* [in Korean] (Ph.D. dissertation, Seoul National University, 1999) pp. 84–103; Ito, *Making Sense of Ryoshiron*, Chapter 3. Student Radicals in Science: Youth Cultures and the Roots of Quantum Physics Research in Late-1920s Japan.

20 Jun Ishiwara, *Fundamental Problems in Physics* [in Japanese] (Tokyo: Iwanami Shoten, 1923).

21 Katsuki, "Before the Returning-Home of Nishina," 752.

22 Ibid., 752–753.

23 Yukawa, *Tabibito*, pp. 160–161.

24 Tomonaga, "My Teachers and My Friends," p. 196.

25 Yoichi Uchida, "The Early Studies of Quantum Mechanics in Kyoto" [in Japanese], *Butsuri*, *45* (1990), 758–760 on 758.

26 Katsuki, "Before the Returning-Home of Nishina," 754.

27 *The History of Physics in Japan*, Vol. 1, p. 302. See also Katsuki, "Before the Returning-Home of Nishina," 753.

28 *The History of Physics in Japan*, Vol. 1, p. 303; Vol. 2, pp. 308–314.

29 Jiro Sasaki et al., "Days of Riken: From the late Taisho to the End of World War II [in Japanese]," *Shizen* (December, 1978), 20–27 on 25.

30 Katsuki, "Before the Returning-Home of Nishina," 753.

31 Ito, *Making Sense of Ryoshiron*, Chapter 3.

32 *Readings in Physics* [in Japanese], Vol. 1 (Tokyo: Iwanami Shoten, 1927). For the analysis of the book, see Ito, Section 14 of Chapter 3.

33 Sasaki et al., "Days of Riken," 25–26.

34 As stated by the president of the Mathematico-Physics Society, cited in *The History of Physics in Japan*, Vol. 2, p. 305.

35 For the full Table of Contents, see *The History of Physics in Japan*, Vol. 2, p. 306.

36 Katsuki, "Before the Returning-Home of Nishina," 754.

37 Helge Kragh, *Quantum Generations: A History of Physics in the Twentieth Century* (Princeton: Princeton University Press, 1999).

38 Sin-itiro Tomonaga, et al., "Remembering Nishina sensei," in *Collected Essays of Sin-itiro Tomonaga*, Vol. 6 [in Japanese] (Tokyo: Misuzu Shobo, 1982), pp. 57–93 on p. 58.

39 Y. Nishina to G. Hevesy (April 1, 1929) in *Supplement to Publications*, pp. 6–8 on p. 6.

40 Takeshi Nawa was Nishina's classmate at the Department of Electrical Engineering in Tokyo Imperial University. Nishina and Miss Mie Nawa however hardly knew each other before marriage.

41 Y. Nishina to G. Hevesy (April 1, 1929) in *Supplement to Publications*, pp. 6–7. Italics are added.

42 G. Hevesy to Y. Nishina (April 24, 1929) in *G. Hevesy–Y. Nishina Correspondence*, 1928–1949, pp. 2–3 on p. 2.

43 Y. Nishina to N. Bohr (April 6, 1929) in *Y. Nishina's Correspondence with N. Bohr and Copenhageners*, 1928–1949, pp. 2–3 on p. 2.

44 Y. Nishina, "Quantum Theory and Causality" [in Japanese], *Denkigakkai Zasshi* (December, 1929), 1331–1345.

45 Nishina regularly read his research results at the Riken's semiannual meetings. On November 19, 1929, he read papers on the line spectrum of lithium and on the method of using line spectrum to investigate the structure of atom; on October 29, 1930, he read papers on secondary x-ray tubes, the relation between x-ray characteristic absorption values and the microscopic structure of rare materials, and the absorption value of liquids; on May 27, 1931, he read a paper on the characteristics of secondary x-ray tubes (coworked with Sanae Yoshida) and on the change of frequency of hologenized silver.

46 Y. Nishina, *X-Rays* [in Japanese] (Tokyo: Koritsusha, 1929).

47 Oh, *Nishina Yoshio and the Modern Physics in Japan*, p. 86.

48 Y. Nishina to H. Nagaoka (January 27, 1928), *Butsuri*, *39* (1984), 160–162.

49 Y. Nishina to N. Bohr (April 6, 1929) and (November 25, 1929) in *Y. Nishina's Correspondence with N. Bohr and Copenhageners, 1928–1949*, pp. 2–3 and pp. 13–14.

50 Y. Nishina to P. A. M. Dirac (April 20, 1929) in *P. A. M. Dirac–Y. Nishina Correspondence, 1928–1948* (Tokyo: Nishina Memorial Foundation, 1990), pp. 1–3 on p. 3. Italics are added.

51 Y. Nishina to N. Bohr (November 25, 1929) in *Y. Nishina's Correspondence with N. Bohr and Copenhageners, 1928–1949*, pp. 13–14 on p. 13.

52 Y. Nishina (ed. and trans.), "Heisenberg and Dirac's Lectures on the Problems of Quantum Mechanics," *Keimeikai Gaiyo*, *11* (1932), 3.

53 Y. Nishina, "Dr. Langmuir's Visit to Japan [in Japanese]," *Kagaku*, *5* (1935), 29–32; "Professors Dirac and Beck's Lectures (I) [in Japanese]," *Kagaku*, *5*

(1935), 359–361; "Professors Dirac and Beck's Lectures (II) [in Japanese]," *Kagaku, 5* (1935), 400–402; "Professor Niels Bohr's Visit to Japan [in Japanese]," *Kagaku, 7* (1937), 207–210.

54 Y. Nishina to P. A. M. Dirac (April 20, 1929) in *P. A. M. Dirac–Y. Nishina Correspondence, 1928–1948*, pp. 1–3. See also P. A. M. Dirac to Y. Nishina (December 19, 1928 and May 11, 1929), *P. A. M. Dirac–Y. Nishina Correspondence, 1928–1948*, p. 1 and pp. 4–5; Y. Nishina to P. A. M. Dirac (June 11, 1929), *P. A. M. Dirac–Y. Nishina Correspondence, 1928–1948*, pp. 5–6.

55 Hidehiko Tamaki, "The Window to the World Academy Opened [in Japanese]," in *Nishina Yoshio*, pp. 81–90 on p. 84.

56 Sin-itiro Tomonaga, "The Last Lecture: Addenda and Recollections," in *The Story of Spin*, translated by Takeshi Oka (Chicago: University of Chicago Press, 1997), pp. 215–231 on pp. 222–223.

57 Nishina Archive, MSS 752, 753, 754, 755, 758, 760, 765, 772, 773, 778. These letters from various universities reported the number of copies of the book that were needed.

58 Tamaki, "The Window to the World Academy Opened," pp. 87–88.

59 G. Hevesy to Y. Nishina (April 24, 1929) in *G. Hevesy–Y. Nishina Correspondence, 1928–1949*, pp. 2–3.

60 G. Hevesy to Y. Nishina (January 9, 1930), ibid., pp. 4–5.

61 Y. Nishina, "Summaries of Professor Hevesy's Lectures [in Japanese]," *Bullettin IPCR, 10* (1931), 616–623.

62 Letters between Bohr and Nishina in 1929–1936 demonstrate that Bohr had tried several times to visit Japan during this period. See *Y. Nishina's Correspondence with N. Bohr and Copenhageners*, pp. 13–59.

63 *The History of Physics in Japan*, Vol. 1, p. 306.

64 Y. Nishina, "Professor Niels Bohr's Visit to Japan [in Japanese]," *Kagaku, 7* (1937), 207–210; "Atom," *Kagakuchishiki* (July, 1937), 10–17.

65 Ito, *Making Sense of Ryoshiron*, Chapter 7.

66 Y. Nishina to G. Hevesy (April 1, 1929) in *G. Hevesy–Y. Nishina Correspondence, 1928–1949*, pp. 6–8 on p. 6.

67 Nishina was a part-time lecturer at Tokyo Imperial University from October, 1937 to December, 1942. See *The One Hundred Year History of Tokyo University: History of Departments* [in Japanese], p. 249.

68 Y. Uchida, "The Early Studies of Quantum Mechanics in Kyoto," 759.

69 Y. Sugiura, "Über die numerische Bestimmung der Mittelwerte zwischen Ortho- und Paratermen von He and Li+ bei Berücksichtigung des Polarisationsgliedes in der quantenmechanischen Störungstheorie," *Zeitschrift für Physik, 44* (1927), 190–220; "Über die Eigenschaften des Wasserstoffmoleküls im Grundzustande," *Zeitschrift für Physik, 45* (1927), 484–492; "Application of Schrödinger's Wave Functions to the Calculation of Transition Probabilities for the Principal Series of Sodium," *Philosophical Magazine, 4* (1927), 495–504.

70 Ito, *Making Sense of Ryoshiron*, Chapter 4, p. 174.

71 Tomonaga, "The Last Lecture," pp. 225–226.

72 Ibid., p. 226.

73 Tomonaga, "Nishina sensei's Warm Heart Moved Me [in Japanese]," *Shizen* (March, 1971), 262–265 on 264.

74 Tomonaga, "The Last Lecture," p. 226.

75 Ibid.

76 Tomonaga, "Nishina Sensei's Warm Heart Moved Me."

77 Y. Nishina to N. Bohr (April 6, 1929), *Y. Nishina's Correspondence with N. Bohr and Copenhageners, 1928–1949*, p. 2; Y. Nishina to N. Bohr (August 23, 1930), *Supplement to the Publications*, p. 10.

78 Ezawa and Takeuchi, "The Chronology of Yoshio Nishina," pp. 278–279, and Seiji Kaya to Y. Nishina (October 6, 1931), Nishina Archive, MSS 763.

4 Beloved *Sensei*: Theoretical Research and the Emergence of a Research Network

Dr. Nishina was a theoretician, but he had a good feeling for experiments. His ability, though, was that of a theorist.

Sin-itiro Tomonaga[1]

I was a very lucky man. I regarded Nishina-*sensei* as my father, and have had Tomonaga-san as a comrade ever since high school years.

Hideki Yukawa[2]

4.1 A NEW KIND OF BOSS AT RIKEN

The establishment of the Nishina Laboratory at Riken opened a new era for the Japanese physics community. All of the four major research fields studied there — quantum mechanics, cosmic rays, nuclear physics, and radioactive biology — were relatively new. Unlike most Japanese physicists who were satisfied to follow in the footsteps of great Western physicists, Nishina intended to compete with Westerners. He carefully focused on pioneering fields of interest and then secured the necessary financial support to establish a series of research groups, starting with those having the lowest start-up costs, the theory and cosmic ray groups. Nishina initiated both of these groups with the opening of the Laboratory, but, as discussed in the following chapter, the cosmic ray group produced important results only after securing necessary funding from the 10th subcommittee of the Japan Society for the Promotion of Science. The nuclear research and radioactive biology groups, which will be discussed in Chapter 6, took shape after 1937 with the completion of a cyclotron.

The opening of the Nishina Laboratory happily coincided with the improvement in Riken's financial situation, which provided great impetus for the Laboratory's growth. In "Paradise for Scientists," Tomonaga reminisced about Riken's atmosphere of quiet affluence during the 1930s:

> Money was of no concern there. When we needed some things for research, if they were not so big or so expensive, we just went to the storage room of Riken, asked for

> them, and signed the necessary documents. That was the only requirement. Since I was not an experimentalist but a theoretician, I just needed some notebooks, pencils and papers for my research that the storage room could supply without any restriction. It was a very nice system.
>
> I soon realized that that was just the tip of the iceberg. The Nishina Laboratory was notorious within Riken because its balance sheet was always deep in red ink. Of course, Nishina did not waste the allocated budget. Since all research there was new, the researchers could not predict the results correctly. This unpredictability resulted in a balance sheet deep in the red. However, the top management silently took care of it. Some laboratories in Riken often made it into the black but their budgets were not cut in the next year. So, those laboratories did not waste money buying unnecessary equipment in order to spend their allocated budgets at the end of the fiscal year as others often did.[3]

Distinguishing the Nishina Laboratory from others in Japan were not only its new lines of research and solid financial base but also its uniquely congenial atmosphere. From the start, Nishina was a very different kind of boss. He would discuss any subject with the young researchers under him and, unlike most Japanese laboratory directors or professors, he was never authoritarian in his manner when meeting with or guiding his subordinates. When Tomonaga joined the Laboratory, he expected to encounter the traditional Japanese academic atmosphere of authoritarian constraint, characterized by strict obedience and hard work, but he was pleasantly surprised. "What surprised me most," he wrote, "was the free atmosphere. This was true not only for the Nishina Laboratory but also for Riken as a whole. Everyone was relaxed and everything was comfortable."[4]

In addition, Nishina never hesitated to delegate full responsibility and authority to capable young scientists, and they enjoyed and appreciated the privilege of being able to approach Nishina to discuss their work. Discussing a project with Nishina, however, carried the risk that, in his enthusiasm, Nishina would recommend projects that were too complex to take on. Hidehiko Tamaki, who entered the Nishina Laboratory in 1934, told a story he thought could be called, "Tomonaga's Method of Boss Management":

> At one time Tomonaga advised me that even if "the Boss" (as we called Dr. Nishina) suggested some outrageously time-consuming and difficult calculations, I should not say it was impossible on the spot. Dr. Nishina, who had studied under Professor Bohr, regarded discussion as an important instrument in research, and Tomonaga was the best possible discussion partner for him. Once when they were in the library, they became so engrossed in discussion of a paper published in a newly arrived journal that they forgot where they were and were shushed by the other researchers around them. Another discussion that they began while standing in a corridor continued on well past suppertime. The Boss's line of argument would often be, "If this doesn't work, then, how about that?" As a result, outrageously complicated calculations sometimes came up in discussion. In those days, we thought about everything in terms of perturbation theory, and "complicated calculation" meant higher-order perturbation. Tomonaga would never say "no" to Dr. Nishina right away. He would just answer, "I shall think about it." Then he would work up an estimate and report to the Boss later that the calculation would require so many tens of thousands of pieces of calculation paper and take so many years. The Boss could then be persuaded to drop the matter.[5]

Minoru Kobayasi, another 1934 entrant to the Nishina Laboratory, remembered the supportive atmosphere at the Laboratory, which he attributed to Nishina's influence:

> Under Dr. Nishina, Tomonaga-san led the theoretical physics group and Dr. Sagane the experimental physics group. The word "organization" gives an impression of formality, but the whole Laboratory was really a very easygoing, free and friendly group. We related to each other as equals, except in research, and the senior members were kind and solicitous toward newcomers like me. I felt then that this highly supportive atmosphere, exceptional especially in those days, was a naturally occurring phenomenon. However, in hindsight, I realize that it was probably a reflection of Dr. Nishina's personal ambitions. I think that he was eager to create the kind of atmosphere that he had experienced during his 8-year stay in Europe, in the circle of highly talented men who followed and revered Professor Niels Bohr. . . .
>
> As we spent time with Dr. Nishina at the Laboratory, we gradually became aware of the depth of his respect for Professor Bohr, and came to assume that Dr. Nishina had tried to model his own Laboratory after Professor Bohr's. Since there was then no other research institute in our country which could have possibly realized such an ideal, it would not be too much to say that Dr. Nishina was lucky to have established his base of operations at the Institute of Physical and Chemical Research. I must also say that we were very lucky to have been part of it all.[6]

Many other Japanese physicists and historians of science believe that Nishina consciously reproduced in his Laboratory the nonauthoritarian atmosphere of Niels Bohr's Copenhagen Institute. "Nishina," wrote the editors of a special supplement to the *Progress of Theoretical Physics* that celebrated Tomonaga's 60th birthday, "brought back with him the 'Kopenhagener Geist' which introduced a breath of fresh air to physicists in Japan."[7] Takeo Hori, reflecting on Bohr's Group in *Nishina Yoshio*, described the "Copenhagen spirit" as a "unique spirit supported by a cooperative milieu, free discussion without any formality, and humor."[8] Interestingly, whereas Japanese writers used the term *Copenhagen spirit* to describe the milieu of Bohr's Institute, Heisenberg and other Westerners used it to describe a preference for their own interpretation of quantum physics, that is, Bohr's complementarity.[9] From the point of view of Bohr's group, Nishina brought to his laboratory *Copenhagen funiki* (the Copenhagen atmosphere), not *Copenhagen seisin* (the Copenhagen spirit).

In short, Nishina equipped his Laboratory with the dominant features of the most influential physics laboratories of the late-nineteenth and early-twentieth centuries, such as the Cavendish Laboratory and Bohr's Copenhagen Institute: pioneering research subjects, an atmosphere of freedom, and great leadership. Nishina would use these resources to lead young Japanese physicists in several unexpected but very successful directions.

4.2 THE THEORY GROUP

Theoretical research in the new quantum mechanics began at the Laboratory with Tomonaga's recruitment from Kyoto in April of 1932, a year that Tomonaga later remembered as "quite meaningful":

> A letter from Professor [*sic*] Nishina arrived at the beginning of 1932. In this letter he asked whether I wanted to study in his lab at RIKEN. I was very hesitant, and I had reservations, not being sure whether I could live up to the expectations of those world-famous professors at a top-notch place like RIKEN, but I also hoped that this would be a good chance and [thought] I should jump at the opportunity. I wrote these feelings honestly to Professor [*sic*] Nishina, to which he replied, "Why don't you come for two or three months to try it. You can continue that calculation on Na_2, or there are many other interesting projects, so you can come and decide." Thus I went to Tokyo at the end of April, 1932 and became a member of Nishina's lab in RIKEN.[10]

Nishina trained Tomonaga thoroughly and rigorously. In 1933, Akira Shimizu, a middle school student and the future secretary-general of Japan's Film Library Council, met Tomonaga at a summer retreat where Tomonaga worked on calculations every day with Nishina and Sakata. When Shimizu asked Tomonaga, "Why don't you goof off once in a while?" Tomonaga replied, "I would very much like to, but the old man [Nishina] will not release me."[11]

Earlier that spring, a small theory group emerged in the Laboratory, when Sakata, having just graduated from Kyoto Imperial University, joined with Tomonaga to carry out the calculation for creation of a positive and negative electron pair by γ-ray in the Coulomb field of an atomic nucleus. They worked together for a year, then, when Sakata moved to Osaka Imperial University in 1934, he recommended as his replacement a former classmate at Kyoto Imperial University, Minoru Kobayasi. Almost at the same time, Tamaki, a graduate of Tokyo Imperial University, joined Tomonaga and Kobayasi, completing the core of the theory group during the 1930s. Under Tomonaga's leadership, the three researchers formed into two pairs — Tomonaga–Kobayasi and Tomonaga–Tamaki — to carry out calculations suggested by Nishina. This cooperative effort duplicated the Klein–Nishina collaboration in calculating Compton scattering at Bohr's Copenhagen Institute some years before.

The theoretical research conducted in the Nishina Laboratory from 1931 to 1945 consisted of four distinct periods. The first of these was research conducted by Nishina with the assistance of his junior researchers from 1931 to 1936. The second was translation into Japanese of Dirac's book, *The Principle of Quantum Mechanics*. The third was work performed by Tomonaga and other researchers from 1936 to 1939. The fourth, which lasted from 1940 to 1945, when Tomonaga's leadership became more evident, was work focusing on meson theory.

4.2.1 Theoretical Works by Nishina and His Junior Researchers

Nishina's first theoretical investigation focused on the creation of positive and negative electrons by photons. The "first work published from our Laboratory" was a short note sent by Nishina and Tomonaga to the *Proceedings of the Physico-Mathematical Society of Japan* that explained how to understand this phenomenon "according to Dirac's theory of the anti-electron."[12] In this note, Nishina and Tomonaga argued that the range of the positive electron offered by P. M. S. Blackett and G. Occhialini

FIGURE 4.1 Sin-itiro Tomonaga (left) was the de facto leader of the theory group in the Nishina Laboratory. The right photograph shows some members of the theory group. Front row from the left: Minoru Kobayasi, Kotaro Tomiyama, Hidehiko Tamaki. Back row from the left: Kanetaka Ariyama, Sin-itiro Tomonaga, Toshinosuke Muto. (Courtesy of the Institute of Physical and Chemical Research.)

should be revised to take into account Dirac's theory as the creation of pairs of positive and negative electrons by means of charged material particles also was possible. They did not provide any detailed calculations to support their argument, however.

When Sakata arrived at the Laboratory from Kyoto in the spring of 1933, he quickly enriched their research on this topic. Tomonaga later recalled the theory group's single mindedness in completing the calculation:

> We decided to give up summer vacation and to push ahead with the research as quickly as possible . . . . [I]t was the summer of 1933. There was an in-service training institute or rather YMCA dormitory called Tôzan-sô in Gotemba near Mt. Fuji. The three of us, Dr. Nishina, Sakata and I confined ourselves there and set to work.[13]

Their result was first published in Japanese in the September and October issues of the *Kagaku* in 1933, and the full paper was read at the semi-annual meeting of Riken on November 17, 1933.[14] Just before Nishina's team finished writing the English version, however, results of studies on the same subject were published by Guido Beck in the *Zeitschrift für Physik* and by Robert Oppenheimer and M. S. Plesset in the *Physical Review*.[15] In December 1933, W. Heitler and F. Sauter published a paper in *Nature* on the same subject that contained the full calculation.[16] The result of the Nishina–Tomonaga–Sakata calculations for the cross section of the positive electrons was slightly different from Oppenheimer and Plessest's but generally agreed with that of Heitler and Sauter. Nishina, Tomonaga and Sakata's paper, "On the Photo-Electric Creation of Positive and Negative Electrons," finally appeared in May of 1934.[17] At about the same time, Hans Bethe and Heitler published "more complete calculations on this problem" in the *Proceedings of the Royal Society*.[18]

Next, in 1934, Nishina and Tomonaga wrote a paper on the negative-energy electron.[19] In it they explained that Dirac's basic assumption that the electrons filling up the negative-energy state are observable was the result of "the exclusion principle of Pauli together with the laws of the conservation of energy and momentum."[20] Nishina and Tomonaga also supported the view that:

> The proton may be considered to be a neutron with a hole in the state of negative kinetic energy, i.e. a close combination of a neutron and a positive electron, as has been pointed out by different authors. Any nucleus can be then looked upon as a combination of neutrons and holes.... From this consideration it follows that there is only one kind of electricity, i.e. the negative charge of the electron, the positive being merely the hole in the closed sates filled up with electrons of positive or negative kinetic energy according to Pauli's principle.[21]

Also in 1934, Nishina, Tomonaga and Tamaki published a paper on the cross section for the creation of a photon in "the annihilation by recombination of an incident positron and a K-, L_I-, L_{II}-, L_{III}-electron belonging to a bare atomic nucleus of charge *Ze* at rest."[22] Unlike Fermi and Uhlenbeck's previous work on the same topic, which used nonrelativistic radial eigenfunctions, Nishina's team made its calculations "following our original programme, in a complete relativistic manner." Their results for slow positrons were larger than those obtained by Fermi and Uhlenbeck "by a factor of about two."

In the summer of 1935, Nishina, Tomonaga and Kobayasi finished a paper on the cross sections for the creation of pairs of positive and negative electrons by heavy charged particles of very high velocity on colliding with atomic nuclei.[23] Because the calculations were much more complicated than those associated with the photoelectric creation of the pair, the three researchers employed the approximate method of Weizsäcker and Williams:

> This method is expected to give a tolerable approximation, when at least one of the colliding particles has a velocity close to that of light with respect to the electron in question in the state of negative energy and, moreover, their motions are not much changed during the creation of pairs.
>
> These conditions can usually be satisfied only for a heavy particle of a very high energy in collision with a still heavier nucleus, which cannot then be set in an appreciable motion: a proton, for instance, colliding with a nucleus of a heavy atom.[24]

After painstaking calculations, the three compared their results with those obtained by other physicists: L. Landau and E. Lifshitz; E. J. Williams; Heitler and L. Nordheim; Oppenheimer; and Nordheim.[25] The comparisons proved satisfactory overall.

In 1936, Nishina, Tomonaga, and Tamaki published a paper on the theory of the collision of the neutron and the proton.[26] After reviewing the recent discovery that the cross section for the collision of slow neutrons with protons is much larger than that for fast neutrons, the authors argued that the explanation might be more plausible if "one assumes both Heisenberg's exchange force and Majorana's force coming into play simultaneously."[27] This work also confirmed Bethe's theory on the subject that had been published in an article the previous year.[28] Tomonaga later lamented the belated publication of this important paper, which lost the Laboratory yet another chance to beat Western physicists in a frontier field:

> I was quite elated with this achievement, and Professor [*sic*] Nishina was also satisfied, and our results were reported successively from the annual meeting of the Physico-Mathematical Society of Japan in Sendai, 1933, to the autumn meeting of RIKEN in 1935. However, probably because he was so busy with a variety of experimental work, Professor [*sic*] Nishina put off publishing this paper. While I was agonizing about this, Bethe and Peierls did exactly the same thing and published it. I was extremely upset (I felt as if I were biting my navel), and I was livid with Professor [*sic*] Nishina.[29]

This paper was the last theoretical work in which Nishina was directly involved.

4.2.2 Translation of Dirac's Principles of Quantum Mechanics

An important contribution of the theory group to the 1930s Japanese physics community was their translation of the second edition of Dirac's *The Principles of Quantum Mechanics* (1935). Nishina became greatly interested in translating Dirac's book in the spring of 1931, just a few months after publication of the first edition, and he wrote to Dirac for permission to do so:

> A friend of mine Mr. U. Doi [Uzumi Doi, a physicist working at Nagaoka's laboratory at Riken] here, has proposed [to] me to translate your book on quantum mechanics into Japanese, in joint-name [*sic*] with him. I agree with him that it is a very useful thing for people here to have a translation of such an excellent book, and although I have very little time to spare at present, I might try to undertake the task if you give us your consent. The translation must be very accurate. You cannot of course see if it is so, but I shall take the responsibility for that if I began at all. I should be much obliged if you would let me know what you think of the matter.[30]

In his reply to Nishina dated April 22, Dirac gave his blessing to the project. "If you decide to proceed with the translation," he wrote, "I should be glad to send you a list of misprints and other corrections, which have been pointed out to me by various people."[31] Dirac also introduced Nishina to the Clarendon Press to facilitate the legal permission, which soon was granted to Nishina and the Japanese publisher, Iwanami Shoten.

Two sets of delays slowed the translation of Dirac's book. First, Nishina's new job as director of his own laboratory at Riken kept him very busy. He explained to Dirac, "As I have not much time at present, the translation might not be ready before long, but I hope to finish it as soon as possible" and, later, "progress is very slow."[32] Dirac further delayed the project by advising Nishina to wait for the second edition, which Dirac expected to appear in July 1932.[33] At that point, Nishina virtually stopped the translation project until October 6, 1934, when Dirac finally reported that he had finished writing the second edition and that it was in press. Dirac had rewritten most of the first edition, "in an attempt to make it easier to understand," and added an entire chapter on the quantum theory of fields. "Do you still think a Japanese translation is desirable?" Dirac asked Nishina. "If so," he suggested, "the present time would be the best time for it."[34] The second edition and permission to translate it arrived at the Laboratory from Clarendon Press early in April of 1935.

Nishina resumed the translation project after Dirac's second visit to Japan in June and July of 1935. The entire theory group, Tomonaga, Kobayasi, and Tamaki, took charge of the translation. They started a special seminar for the project, but progress was very slow. To concentrate on the translation project during the summer, Nishina and his theory group isolated themselves in two borrowed summer houses in Kita-Karuizawa of Naganoharamachi in the Gumma Prefecture. The Nishina family lived in one house and the theory group in the other. Tamaki described their typical work day as follows:

> In the morning we would work up rough translations of the portions allotted to each of us, in the afternoon we discussed and checked our rough translations together, and after supper we would make clean copies of the corrected translation.[35]

Throughout the summer, except for a few excursions around the village and a trip to Tokyo to attend a farewell party for Sagane, who was going abroad, the three young physicists devoted themselves to the translation of the book. By the end of the summer, they "were able to get it into shape by and large."[36] Publication of the translation required yet another year. In February of 1936, Nishina wrote to Dirac, "We have gone through our translation in a preliminary stage but it will take some time before we send it to the press."[37] Nishina also asked Dirac to write a preface to the Japanese edition.[38] After proofreading, which started in May of 1936 and continued until the end of the year, the translation finally appeared in print in December of 1936. Figure 4.2 shows Dirac's preface to the Japanese translation.

Later, in December of 1948, Nishina wrote to Dirac to obtain a copy of the third edition and to request permission to translate it. "Tomonaga and others who translated the former edition into Japanese," he said, "would like to translate the new edition in the same way, if you and your publisher would give us approval."[39] Unfortunately, Nishina did not live to see the translation of the third edition, which was published more than a year after his sudden death on January 10, 1951.

4.2.3 The Works of Tomonaga and Other Researchers from 1936 to 1939

From the mid-1930s onward, Nishina's interest moved to cosmic ray research and the construction of the cyclotron, and Tomonaga became the de facto leader of the theory group. The group remained small: Tomonaga, Kobayasi, and Tamaki were the only permanent members, but a few other researchers often collaborated with them. As Nishina's participation diminished and Tomonaga's influence grew, the group's focus shifted away from Nishina's preference for theoretical interpretations towards Tomonaga's preference for accurate and extensive calculations. Tomonaga was the driving force behind this calculation effort.

In 1937, Tomonaga, Kwai Umeda and Yoro Ono carried out calculations on the exchange and van der Waals forces between two deuterons.[40] During these

THE
PRINCIPLES
OF
QUANTUM MECHANICS
BY
P. A. M. DIRAC
FELLOW OF ST. JOHN'S COLLEGE
CAMBRIDGE

SECOND EDITION

ディラック

量 子 力 學

仁科芳雄
朝永振一郎
小林稔
玉木英彦
共譯

1936

岩波書店

PREFACE TO JAPANESE EDITION

During my visits to Japan in 1929 and 1935 I have several times had the pleasure of lecturing to a Japanese audience on questions connected with the modern quantum mechanics. I am now very glad to be able to present a general account of the subject to a wider circle of Japanese students through the appearance of a Japanese translation of my book.

Cambridge, March 1936.

P. A. M. Dirac.

目 次

I. 重 疊 原 理

II. 狀態とオブザーヴァブル

III. 離散固有値に對する表示理論

FIGURE 4.2 Above: The title page of the Japanese translation of P. A. M. Dirac's *The Principles of Quantum Mechanics* (second edition). Below: Dirac's preface to the Japanese translation is shown in the left, and the first page of the table of contents is shown on the right.

calculations, they found a peculiar characteristic of the overlapping integral that Tomonaga and Umeda focused on more carefully and discussed in their next paper.[41] Next, Tomonaga calculated the kinetic nuclear energy according to the Hartree–Fock model with a new method in which the coordinates of the center of gravity were eliminated from the wave functions of the nucleus.[42] A year later, Tomonaga and Tamaki published the result of their calculations on the collision of a high energy neutrino with a neutron in cosmic rays.[43] They found that "the limit of the incident-neutrino energy, beyond which the shower process of Heisenberg's type may become important, must be of the order of 10^{12} eV," not 10^9 eV as Heisenberg had estimated.[44] In 1938, Tomonaga and Kobayasi published their paper on the cross sections for scattering and splitting of photons by the field of atomic nuclei according to nonlinear field theory of Born and Infeld.[45]

Other researchers in the Nishina Laboratory joined Tomonaga's quest for accurate calculations. In 1936, Umeda and Ono published a paper on the calculations of the binding energy of light nuclei, which they based on S. Flügge's atomic model in which the number of protons is not equal to that of neutrons.[46] They also calculated and published, in 1937, the frequencies of vibrations of a nucleus by treating it as a sphere of the Thomas–Fermi fluid.[47] Umeda calculated the amplitude factor for the Fermi matrix, the distribution of nuclear energy levels according to the oscillator model, the partition numerorum, the temperature change of the liquid-drop model of nucleus, and the fluctuation of nuclear excitation energy.[48] Kanetaka Ariyama made calculations for the properties of divalent metals according to the electron theory of metals.[49]

Kobayasi's works were more formal than similar works in the Nishina Laboratory, although they also included heavy calculations. In 1937, Kobayasi calculated the second-order correction of nuclear energy, starting with the Thomas–Fermi approximation for the heavy nucleus, and he also compared the probability of the process where the incident neutron or proton is captured by the nucleus leaving one of its constituent particles excited.[50] The results suggested the inadequacy of treating the problem with the Thomas–Fermi or the Hartree method and also the failure of the method of perturbation in these processes; therefore Kobayasi supported Bohr's nuclear model.

In the same year, Kobayasi and Shoji Ozaki studied the energy loss of fast charged particles by pair creation.[51] Following the method developed by Williams and Weizsäcker, they calculated the order of magnitude of the average energy loss of a charged particle per collision with an atom. They then applied the result to explain cosmic ray showers, concluding that "the greater part of cosmic ray showers found at depths in matter are initiated by those fast positive and negative electrons which are produced through pair creation, by heavy particles contained in the hard component of cosmic rays."[52]

In 1937, Tomonaga sailed to Germany to work under Heisenberg at Leipzig. When Tomonaga asked Heisenberg for suitable research topics, he steered Tomonaga away from Yukawa's meson theory:

> You plan to stay here for two years. I think it would be good for you to treat something having a concrete basis, at least for the first year. I am interested in Yukawa's theory,

> but its foundation does not seem to be very clear. I suggest that you do not work on it, but on some less speculative topic.[53]

The more conservative topic Tomonaga chose was describing Bohr's liquid drop model of the heating of a nucleus when it absorbs a neutron; his findings were published in the *Zeitschrift für Physik* in 1938.[54] Despite his laborious effort, Tomonaga's conclusion was unconvincing: the calculated viscosity of nuclear matter using Fermi's model proved to be too large to accommodate the oscillations anticipated by Bohr.

Tomonaga then turned his attention to a problem arising from a conflict between Yukawa's meson theory and experimental results. The 1937 discovery of a new heavy particle in cosmic rays quickly focused wide attention on the meson theory, which, however, presented theoretical physicists with difficult problems to solve. In a long review of the topic, Heisenberg and Hans Euler noted discrepancies between theoretical predictions and the data from cosmic rays.[55] This stimulated Tomonaga's interest. According to Yukawa's 1935 theory, β-decay occurs in two steps: first, a meson is created while a neutron changes into a proton; then, the meson transforms into an electron and an antineutrino. Tomonaga proposed an alternative explanation: that the meson decays into an electron and an antineutrino through the intermediate state of a nucleon pair. Heisenberg encouraged Tomonaga, saying, "Your model may give better results than the original one." As Tomonaga remembered:

> I found within a week that there was an integral whose result was infinity, and I explained my result to Heisenberg, who agreed. In those days he was working with Euler, applying positron theory to the problem of vacuum polarization. It seemed to me that the divergence I obtained could be treated by whatever procedure worked for the vacuum polarization, but this problem could not be solved without difficulty, and so on — I was totally confused. It was something like a snake trying to swallow its own tail. Now we all know that this problem is solved by the renormalization procedure.[56]

Tomonaga would return to this topic five years later under very different circumstances.

4.2.4 The Works of Tomonaga and Other Researchers from 1940 to 1945

Tomonaga's two-year absence from the Nishina Laboratory was a great blow to the theory group. Nishina, busy constructing cyclotrons, paid less attention to theoretical subjects. "Without Tomonaga-san," said Kobayasi, Tomonaga's close collaborator at Riken, "the Nishina Laboratory seemed deserted, and I went over to Osaka Imperial University, near my hometown, about one year after his departure."[57] Without Tomonaga at the Laboratory, few new theoretical works were completed.

Tomonaga's return from Germany in the fall of 1939 revitalized the theory group. The Laboratory's milieu had changed dramatically. Theoreticians no longer followed in the footsteps of Western physicists or wandered from one subject to another. Under Tomonaga's leadership, they concentrated on one subject: Yukawa's meson theory.

In early 1935, Yukawa published his first full English-language paper, "On the Interaction of Elementary Particles. I." To eliminate the difficulties inherent in using the theories of Heisenberg and Fermi to explain the interaction of elementary particles, Yukawa proposed the existence of something previously unknown:

> a hypothetical quantum which has the elementary charge and the proper mass and which obeys Bose's statistics. The interaction of such a quantum with the heavy particle should be far greater than that with the light particle in order to account for the large interaction of the neutron and the proton as well as the small probability of β-disintegration.
>
> Such quanta, if they ever exist and approach the matter close enough to be absorbed, will deliver their charge and energy to the latter. If, then, the quanta with negative charge come out in excess, the matter will be charged to a negative potential.
>
> These arguments, of course, of merely speculative character, agree with the view that the high speed positive particles in the cosmic rays are generated by the electrostatic field of the earth, which is charged to a negative potential.
>
> The massive quanta may also have some bearing on the shower produced by cosmic rays.[58]

Initially, Yukawa's revolutionary idea attracted almost no attention either in Japan or in the West. Most leading Western physicists, including Bohr, had long detested the notion of a "new" hypothetical particle to explain nuclear force or other atomic phenomena. Undeterred, Yukawa continued elaborating his theory with Sakata, who had moved from Riken to Osaka.

The year 1937 was a turning point in the development of meson theory.[59] Carl D. Anderson and Seth H. Neddermeyer's 1936 and 1937 papers proved the existence of an unknown heavy particle with a mass more than 100 times than that of the electron. J. R. Oppenheimer and R. Serber soon suggested in a note to the *Physical Review* that the newly discovered particle might be the heavy quantum predicted by Yukawa. Experimental results by J. C. Street and E. C. Stevenson and by Nishina's cosmic ray group in 1937 independently gave weight to Oppenheimer and Serber's suggestion. From then on, theoretical physicists considered Yukawa's new hypothetical particle, named the meson, a serious entity. Those who were unconvinced that the newly discovered particle was Yukawa's hypothetical particle, mostly cosmic ray researchers, preferred to use the nomenclature *mesotron*. The two names had been used in very confused way even in Japan, however. Some doubted about the existence of the hypothetical meson continued until 1947, when physicists finally differentiated the pion (π, which interacts strongly with nucleons and which Yukawa's theory had predicted) from a muon (μ, which Anderson and Neddermeyer discovered in 1937 in cosmic rays).[60] The Yukawa group — Yukawa, Sakata, Mitsuo Teketani, and Kobayasi — produced many important results on the meson theory and achieved world recognition.

Independently of Yukawa's theory group in Osaka–Kyoto, Tomonaga's theory group in the Nishina Laboratory published several important papers on the meson theory.[61] In 1940, Tomonaga published two papers on the subject. The first, which he authored with Gentaro Araki, explained how much the Coulomb force of atomic nuclei influences the capture of slow mesons.[62] Tomonaga and Araki calculated "the probability for a meson of an incident energy E, being captured along its path before it is brought to rest" and found that the effect of the Coulomb force is "very small for

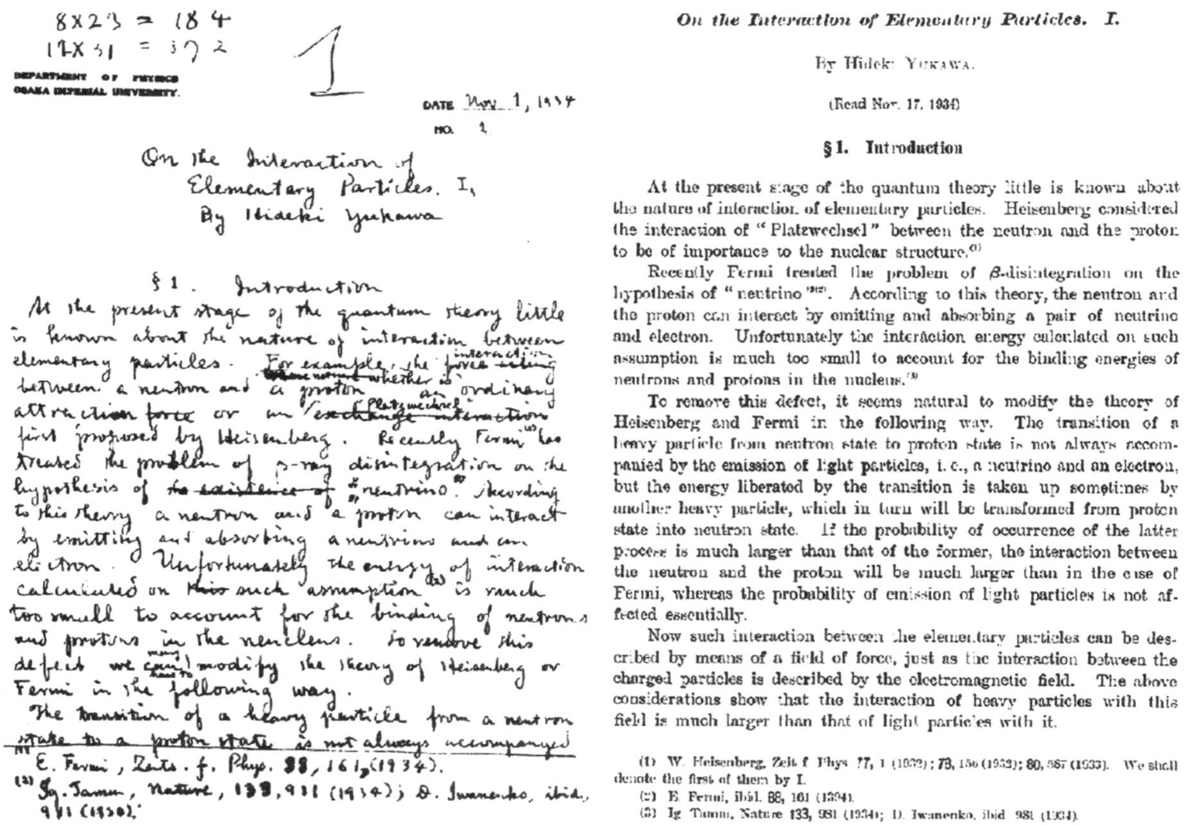

8×23 = 184
17×31 = 372

DEPARTMENT OF PHYSICS
OSAKA IMPERIAL UNIVERSITY.

1

DATE Nov. 1, 1934
NO. 1

On the Interaction of
Elementary Particles. I,
By Hideki Yukawa

§1. Introduction

At the present stage of the quantum theory little is known about the nature of interaction between elementary particles. ~~For example, the force acting~~ between a neutron and a proton whether is an ordinary attraction force or an "Platzwechsel" interaction first proposed by Heisenberg. Recently Fermi[(2)] has treated the problem of β-ray disintegration on the hypothesis of "neutrino". According to this theory a neutron and a proton can interact by emitting and absorbing a neutrino and an electron. Unfortunately the energy of interaction calculated on such assumption is much too small to account for the binding of neutrons and protons in the nucleus. To remove this defect we may modify the theory of Heisenberg or Fermi in the following way.

The transition of a heavy particle from a neutron state to a proton state is not always accompanied

(2) E. Fermi, Zeits. f. Phys. 88, 161, (1934).
(3) Ig. Tamm, Nature, 133, 981 (1934); D. Iwanenko, ibid., 981 (1934).

On the Interaction of Elementary Particles. I.

By Hideki YUKAWA.

(Read Nov. 17, 1934)

§1. Introduction

At the present stage of the quantum theory little is known about the nature of interaction of elementary particles. Heisenberg considered the interaction of "Platzwechsel" between the neutron and the proton to be of importance to the nuclear structure.[(1)]

Recently Fermi treated the problem of β-disintegration on the hypothesis of "neutrino"[(2)]. According to this theory, the neutron and the proton can interact by emitting and absorbing a pair of neutrino and electron. Unfortunately the interaction energy calculated on such assumption is much too small to account for the binding energies of neutrons and protons in the nucleus.[(3)]

To remove this defect, it seems natural to modify the theory of Heisenberg and Fermi in the following way. The transition of a heavy particle from neutron state to proton state is not always accompanied by the emission of light particles, i.e., a neutrino and an electron, but the energy liberated by the transition is taken up sometimes by another heavy particle, which in turn will be transformed from proton state into neutron state. If the probability of occurrence of the latter process is much larger than that of the former, the interaction between the neutron and the proton will be much larger than in the case of Fermi, whereas the probability of emission of light particles is not affected essentially.

Now such interaction between the elementary particles can be described by means of a field of force, just as the interaction between the charged particles is described by the electromagnetic field. The above considerations show that the interaction of heavy particles with this field is much larger than that of light particles with it.

(1) W. Heisenberg, Zeit. f. Phys. 77, 1 (1932); 78, 156 (1932); 80, 587 (1933). We shall denote the first of them by I.
(2) E. Fermi, ibid. 88, 161 (1934).
(3) Ig. Tamm, Nature 133, 981 (1934); D. Iwanenko, ibid. 981 (1934).

FIGURE 4.3 The draft of Yukawa's paper on the meson theory, (left), and the published one (right) in the *Proceedings of the Physico-Mathematical Society of Japan*, *17* (1935), 48–57 on 48. (Courtesy of the Yukawa Hall Archival Library, Kyoto University.)

slow negative mesons, and still less for positives." In addition:

> The negative mesons will be much more likely captured by nuclei than to disintegrate spontaneously, not only in dense materials but also in gases. On the other hand, practically all positive mesons will disintegrate spontaneously because of the extremely small capture probability due to the existence of the potential barrier.[63]

In his second paper of 1940, Tomonaga examined the validity of E. J. Williams' impact parameter method for calculating the average number of electrons when they collide with mesotrons in cosmic rays due to the knock-on process and also the number above a certain level of energy.[64] Tomonaga seriously considered the importance of mesotron spin in this collision, asking how one could adapt Williams' method to a case in which a change takes place in the spin direction. His solution was to replace the point electric charge with an electric dipole, and then apply Williams' method to the ionization of atoms and the production of electron–positron pairs by mesotrons.

In 1941 and 1942, Tomonaga published two papers, *Zur Theorie des Mesotrons, I* and *II*, that extended and generalized G. Wentzel's strong coupling method into a more practical "intermediate coupling method."[65] In the 1941 paper, Tomonaga summarized his work as follows:

> In the present work, it will be shown how the behaviour of a mesotron emitted virtually from a proton or neutron can be investigated by the introduction of the Hartree approximation, not only in the limits of strong or weak coupling but also in cases lying between them. The result, an approximation formula for the self-energy of the nuclear particles, agrees exactly for weak or strong coupling, respectively, with the results of the perturbation method or the method of Wentzel.[66]

The outbreak of the Pacific War in December 7, 1941 had little impact on Tomonaga and other theoreticians in the Nishina Laboratory, at least in the beginning. In 1942, Tomonaga and Tatuoki Miyazima refined this method by introducing more variational parameters and used it to calculate multiple production of mesotrons by the collision of a nuclear particle (nucleon) with a heavy particle (nucleus). Also in 1942, Tomonaga published another paper on the scattering of mesotrons by nucleons, showing that the difficulty of calculating the cross section in this case could be at least partially removed by considering the damping reaction of the meson field on the scattering nuclear particle.[67]

Tomonaga and Miyazima's 1943 paper, "On the Mesotron Theory of the Nuclear Forces," was the culmination of Tomonaga's work on the meson theory before 1945.[68] In it, Tomonaga correctly pointed out the source of certain difficulties in understanding interactions between the forces under study:

> Although this perturbation method is very natural and intuitive and was successfully used to deduce the electromagnetic interaction which results from the exchange of photons between electrons, there are also many reasons to believe that the same method will not be applicable to the case of the nuclear forces.
>
> As has been remarked by many authors, the main reason is that the interaction between the mesotron field and the nucleons cannot be considered small. For the problem of

> the interaction of mesotrons with the nucleons, it will thus be essential to take into account the reaction of the field, which is entirely neglected in the perturbation-theoretical treatment of the problem.... In the present stage of the relativistic quantum mechanics, however, the well known divergency difficulties prevent us from treating such problems beyond the scope of the perturbation theory in a consistent way. The field equations of the present theory must ultimately be regarded, strictly speaking, as having no finite solutions at all.
>
> It may be, nevertheless, expected that, if we interpret the present theory correctly, it will be a good approximation to the forthcoming theory, and to each solution of the fundamental equation of the latter theory there will exist in some way the corresponding solution in the present theory. If we admit this idea the task will be to treat the problem most exactly in this sense according to the present theory and compare the result with the experiment. Our calculations have their meaning always only in this restricted sense.[69]

The authors then adopted the classical approximation method to provide the desired "comparatively powerful means of investigating the influence of the reaction of the mesotron field on the nucleons in the case of large coupling."[70] Silvan S. Schweber, in his *QED and the Men Who Made It*, argued that this methodology reveals Tomonaga's "conservative stance" in believing "that advances were made by limited, incremental, evolutionary steps."[71] Tomonaga wrote two short notes on the interaction between mesons and nucleons during the summer of 1943 but could not expand them into full papers because of the war.[72]

Other theoreticians at the Nishina Laboratory also worked on the meson theory. Kobayasi and Utiyama adopted Weizsäcker and Williams' method of impact parameters to calculate the cross section for the pair creation of mesons by γ-rays.[73] They used the resulting formula to estimate "the differential cross section for 'Bremsstrahlung', that is, the probability for the process in which a meson of very high energy E ($E \gg \mu c^2$) collides with an atomic nucleus and emits a photon." Araki published two papers on the meson theory: one focused on the production and capture of the meson of a spin 0, pseudoscalar meson, and the other focused on the nuclear scattering of pseudoscalar mesons.[74] Miyazima examined to what extent the theory of "the 'mesotron' of spin 1/2" works for the interaction of the meson with heavy and light particles, nuclear forces, life-time of the meson and the β-decay, and the collision of mesons with heavy particles.[75] Ozaki attempted to estimate the effects of spin on the scattering of mesons by the electrostatic field with spherical well potential "by solving the wave equation exactly."[76] Tamaki published a paper in 1942 on the spin of the meson.[77]

The few papers that did not focus on the meson theory included Ariyama's 1941 paper on the theory of superconductivity and Tamaki's 1942 paper on the relativistic equations for particles of arbitrary spin in an electromagnetic field.[78] The most notable of these, surprisingly, was written by Tomonaga himself, "On a Relativistically Invariant Formulation of the Quantum Theory of Wave Fields," which was published in 1943 in the *Bulletin of the Institute of Physical and Chemical Research*.[79] After the end of the war, the idea Tomonaga presented in this article evolved into his famous renormalization theory.

As Japan started losing the war from 1943, most scientists in the Nishina Laboratory became deeply involved in various war-related projects such as the developments of radar or nuclear bomb. Tomonaga and Masao Kotani of Tokyo Imperial University, for example, were developing a theory of the magnetron, a key device with which the radar system could generate electromagnetic waves.[80]

In conclusion, Nishina's contributions to the theory group between 1931 and 1945 were important but limited and indirect. He was the magnet that attracted talented young theoreticians like Tomonaga, Sakata, Kobayasi, and Tamaki into his laboratory at Riken. Under his guidance, a circle of young physicists developed into Japan's first true theoretical research group. In the beginning of the group's formation, Nishina selected these young physicists' research topics, including the production of positive and negative electrons, through which they received thorough training. Assigning this group the task of translating Dirac's *Principles of Quantum Mechanics* into Japanese was one example of Nishina's method of training the theoreticians working under him. The theory group in the Nishina Laboratory advanced so quickly that, as early as in 1936, Tomonaga could manage it without Nishina's supervision. However, the group still needed Nishina's advice, discussion and, most of all, encouragement.

4.3 THE EMERGENCE OF A RESEARCH NETWORK

When the success of Yukawa's meson theory became evident in 1937, physicists in both Japan and the West began to take it seriously. This success could have become yet another episode for a Japanese scientist, as Nagaoka's Saturnian ring model of the atom had been in the first decade of the twentieth century or Ishiwara's quantum mechanics had been in the 1910s. However, this time the emergence of a very efficient research network during the 1930s played a pivotal role in keeping Japanese research on the world physics scene. A group of Japanese physicists in Tokyo, Osaka, and Kyoto continued to produce top-quality research papers on meson theory throughout the 1930s and 1940s because, unlike their predecessors, they no longer worked separately but freely exchanged information.

At first, the network was small. In Kyoto and Osaka, Yukawa formed a small theory group in which Sakata, Taketani, and Kobayasi were his major collaborators.[81] Meanwhile, at Osaka Imperial University, Seishi Kikuchi was busy constructing and operating accelerators with H. Aoki, K. Husimi, Y. Watase, and J. Itoh, who comprised a strong nuclear physics group. These two groups were very close and exchanged ideas. The Kikuchi group often cited Yukawa and Sakata "for valuable discussion"; in addition, Aoki employed Yukawa and Sakata's calculation to produce the "efficiency of the search counter in the case of γ-ray emitted from Cu."[82] Reciprocally, Yukawa and Sakata frequently used experimental data gathered by Kikuchi group.[83] One important exchange of information occurred when Husimi provided Yukawa and Sakata with hard-to-get copies of papers by Fermi and Majorana:

> If I have made any contribution to the meson theory of Yukawa, the most important is that I gave to Yukawa and Sakata the papers of Fermi and Majorana. At that time only

> Tokyo [Imperial] University got the Italian journal, not Kyoto or Osaka. So I brought the papers of Fermi and Majorana in *Ricerca Scientifica*.
>
> It was customary for the Kikuchi Laboratory and Yukawa Laboratory to take lunch together every day. At those lunches we discussed all sorts of questions including political matters. I think I heard Yukawa's first lecture on the meson theory there.... I remembered that I interpreted the Yukawa theory as that of a kind of photon with mass. So I nicknamed the new particle as "heavy quantum," as opposed to the ordinary light quantum.[84]

Yukawa later used the term "heavy quantum" that Husimi coined at this lunch gathering as the term for meson in his 1935 paper. In his autobiography, *Tabibito (The Traveler)*, Yukawa credited the Kikuchi group for the helping with the birth of the meson theory:

> I spoke to everyone about the new theory during the meeting of the Kikuchi research group. Kikuchi said, "If there is such a charged particle, it should become visible in the Wilson cloud chamber, should it not?" I answered, "Yes, the particle can be found in the cosmic ray."[85]

Yukawa's move to Kyoto Imperial University in 1939 to succeed his former mentor, Kajuro Tamaki, provided the group with opportunities for expansion.

FIGURE 4.4 Early days of the Kyoto–Osaka group. Seishi Kikuchi's experimental team and Yukawa's theory team worked together. Yukawa stood in the first from the left in the back row, while Kikuchi sat in the center in the front row. (Courtesy of the Yukawa Hall Archival Library, Kyoto University.)

In Tokyo, the theory group in the Nishina Laboratory was the core of the research network. Moreover, as the next chapter will indicate, the close relationships between the Laboratory's theorists and experimentalists enabled them to freely exchange ideas and data and even sometimes to ask others to do specific work. The general atmosphere of the Nishina Laboratory was "full of freshness," as Tomonaga later remembered:

> All the members were young; even our great chief Nishina was still in his early forties. We all got together after lunch everyday, an eager group of people discussing various matters, not only physics but also such things as plans for beer parties, excursions and so on.[86]

The enthusiasm of the young researchers of the Nishina Laboratory and its counterpart in Osaka no doubt helped them overcome the traditional rivalry between the two regions of Kanto (Tokyo) and Kansai (Kyoto–Osaka) to form a single, highly efficient, research network. Members of the network visited each other, attended each other's seminars, and met jointly in the "meson club" to discuss recent development of meson theory. They transferred easily from one group to the other in an intertwining pattern that was constantly changing. For example, Sakata and Kobayasi were Yukawa's first students at Kyoto Imperial University. After graduating in 1933, Sakata moved to Riken to work with Tomonaga, while Kobayasi remained in Kyoto to work with Yukawa. In 1934, they changed places: Sakata returned to Yukawa in Osaka while Kobayasi joined Tomonaga at Riken (and then, four years later, returned to Yukawa). Exchange of this kind was very unusual in the Japanese academic environment at that time when the loyalty to the professor and the academic factionalism were the rule.

One exceptional figure, Mitsuo Taketani, whom many described as the driving force of the meson-club, transferred from one group to the other for political reasons. Taketani was not only a 1934 graduate of Kyoto Imperial University and a close collaborator of Yukawa and Sakata, but also an important member of an underground antimilitarist circle in the late 1930s, which got him into serious trouble.[87] In June of 1938, while he was collaborating with Yukawa and Sakata on a paper about the meson theory, Taketani was arrested for his underground activities; the following April he was imprisoned. Fortunately, his prosecutor was impressed by his work and suspended Taketani's prosecution. The prosecutor "summoned Yukawa" and released Taketani from prison into his custody. Taketani, who no longer could live and work freely in Kyoto, moved to the Nishina Laboratory in Tokyo, where he soon became a prominent member of the theory group.

A stream of acknowledgments in various papers pointed to the ongoing communications throughout the research network. Researchers from the Osaka–Kyoto group acknowledged the contributions of Nishina and Tomonaga in Tokyo, who reciprocally thanked Yukawa and his colleagues in Osaka–Kyoto. In Yukawa's famous 1935 paper, "On the Interaction of Elementary Particles. I," he employed Tomonaga's calculation of "the mass defect of H^2 and the probability of scattering of a neutron by a proton," acknowledging that he owed "much" to his former classmate.[88] In acknowledging input into their research on Yukawa's meson theory, Kobayasi and T. Okayama in Osaka, thanked both Yukawa and Nishina "for encouragement and discussions."[89] Taketani and Sakata concluded their paper on the meson theory with "cordial thanks" to Yukawa, Tomonaga, Kusaka, and Sakata, also for "encouragement and discussions."[90] Kobayasi and Utiyama, in their paper on the meson theory, expressed gratitude to Nishina and Yukawa for "kind interest" and also indebtedness to Tomonaga and Taketani "for valuable advice."[91]

Two factors explain the remarkable development of this research network in the traditionalist Japanese academic community of the 1930s and early 1940s: Nishina's leadership and Yukawa's meson theory. It was Nishina who brought the Tokyo group into contact with the Osaka–Kyoto group. Unlike most Japanese senior scientists, he opened his laboratory to all talented young Japanese physicists, without regard to their academic background or regional origin. Promising graduates from Tokyo, Kyoto, Osaka, and Tohoku Imperial Universities, as well as from less prestigious institutions, all were welcomed to work together. As Yukawa pointed out, Nishina's open mindedness and generosity contributed enormously to the productivity of the young researchers in the network.

> His laboratory was very open, so we could really go there to talk with him and his staff about many things and in a very informal way, so that mostly it was quite different from the standard atmosphere of the university.[92]

Nishina had the reputation of being the only physicist respected by all Japanese physicists, whether senior or junior in rank or theoretician or experimentalist in interest. Without Nishina, the two future Japanese Nobel Laureates, Yukawa and Tomonaga, might never have collaborated. They had been academic rivals since their undergraduate years at Kyoto Imperial University, and their personalities and working styles were considerably different (Figure 4.5). Yukawa himself observed that he and Tomonaga were "different from each other in many respects."[93] The seeds of commonality that Yukawa recognized, such as their "similar earlier background" and an ability to recognize their "different points of view" as points of "intellectual stimulus" might never have taken root without Nishina's rare ability to inspire researchers to work together.[94]

Seitaro Nakamura, who had worked under both Yukawa and Tomonaga, was familiar with their different temperaments. Whereas Yukawa assigned questions to his students and required them find their own answers, Tomonaga examined and corrected every calculation of his students. Yukawa loved reading Chinese classics, including the works of Chuangtzu and Mencius, but Tomonaga preferred to attend popular shows and to read Japanese traditional fables. When a publisher invited them both to a famous Kyoto restaurant, Yukawa sat upright in the center of the room while Tomonaga leaned against the wall and stretched out his legs.[95]

In contrast to Tomonaga, who followed a step-by-step method and concentrated on eliminating any inconsistencies in theory by performing careful calculations, Yukawa often avoided performing time-consuming calculations by intuitively attacking the core of the problem. Yoichiro Nambu captured Yukawa's style as follows:

> It seems to me that Yukawa's approach had a heuristic and phenomenological tone [that served as a] useful guiding principle to explore uncharted and ever surprising new worlds, just as Lawrence's cyclotron and its descendants served as useful experimental tools to go with it. Yukawa's approach had both conservative and radical sides. He was

FIGURE 4.5 Yukawa and Tomonaga had been academic rivals since their undergraduate years at Kyoto Imperial University. Their personalities and working styles were quite different. Left: J. R. Oppenheimer (left), Yukawa (center), and Tomonaga (right) at the Institute for Advanced Study in Princeton in November 1949. Right: Yukawa and Tomonaga in September 1957. (Courtesy of AIP Emilio Segrè Visual Archives.)

> conservative in the sense that he pursued the logical consequences of relativistic quantum field theory .... But he was radical in the sense that he did not hesitate to speculate on the existence of new elementary particles which had not been seen .... At any rate, Yukawa solved the problem of nuclear forces by dividing it in two inherently different parts: the theoretical framework in which to describe nature, and the substantive question of what entities exist to be described.[96]

It was a function of Nishina's genius that he could connect the theory groups of two such different personalities as Yukawa and Tomonaga without friction.

No doubt Nishina's personal encouragement of Yukawa played as important a role in helping to develop ease of communication in the network as it did in Yukawa's development of the meson theory, a debt that Yukawa never forgot and which he mentioned at every opportunity (Figure 4.6). At the close of his epoch-making 1935 paper, he extended "cordial thanks to Dr. Y. Nishina and Prof. S. Kikuchi for the encouragement throughout the course of the work."[97] In his 1950 article on the birth of meson theory, Yukawa provided a more detailed acknowledgment:

> Dr. Nishina, who was present at [the 1933 Sendai meeting of the Physico-Mathematical Society], suggested that it might be possible to postulate the existence of electrons satisfying Bose statistics. This was the first hint toward the meson theory. From that day on, Dr. Nishina took the most active part in encouraging my investigation.[98]

Yukawa told a similar story in his autobiography: "In November [of 1934], I presented the new theory to the Osaka branch of the Physico-Mathematical Society of Japan. Professor Nishina was very interested in the theory, and he congratulated me."[99] In those early years, Nishina's recognition certainly encouraged Yukawa and helped him develop the confidence he needed to continue perfecting his meson theory.

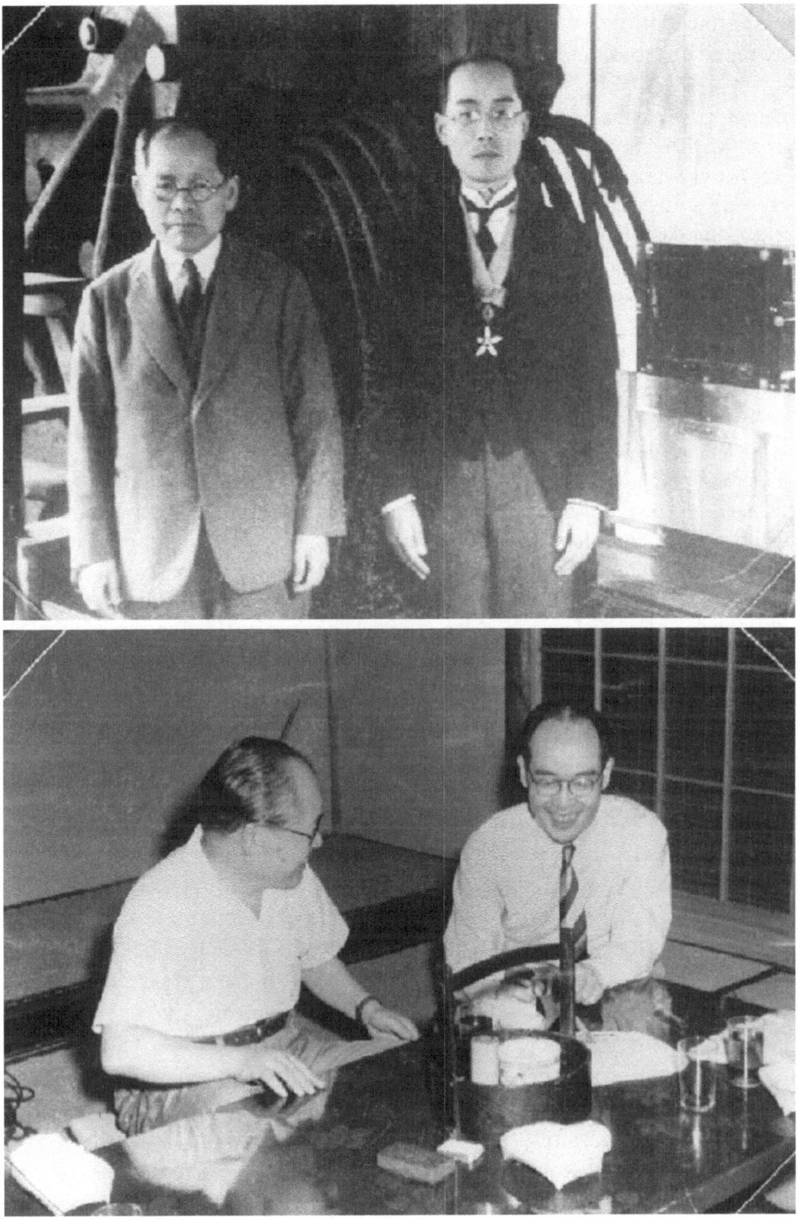

FIGURE 4.6 Nishina's encouragement of Yukawa contributed greatly to the development of meson theory. Yukawa had never forgotten it and acknowledged his personal debt to Nishina. Left: Yukawa visiting Nishina in 1943 after receiving the Decoration of Cultural Merit. Right: Yukawa, 1949 Nobel Prize Winner in physics, visiting Nishina in 1950 after returning from the United States. (Courtesy of the Special Collections, NCSU Libraries.)

Kobayasi's anecdote below illustrates how little support for his meson theory Yukawa initially received from the Japanese physics community:

> Just after he wrote the first paper, Yukawa talked in Tokyo at a regular meeting of the Physico-Mathematical Society of Japan [in April of 1934]. I was in the Research Institute (Riken) at that time and attended the meeting. In those days Yukawa was very shy and spoke very softly. There was no response from the audience except for one question. One person asked if he would repeat his talk from the beginning. The whole audience laughed. I thought the idea was very interesting, but the general impression was that it was only a "neat idea" without much chance to be correct.[100]

In a public speech in 1965, Yukawa once again acknowledged his personal debt to Nishina:

> As I remember now, Nishina-*sensei* played a role of father to me. Needless to refer to Freudian psychoanalysis, all boys regard their own fathers as annoying persons (laughter). My own father was a scholar and also a so-called "thunder-grandfather." I tried not to see him often since whenever he saw me he shouted [at] me (laughter). One of my elder brothers was not afraid of him and spoke [to] him freely. My father got angry at this attitude and said, "The boy doesn't know etiquette." I believe this was quite wrong . . . . But Nishina *sensei* was different. He was very kind and broad-minded. Whenever I told him various ideas [I had] in mind, he said, "They seem very interesting." There was no one who had told me [anything] like that.
>
> Although he was extremely busy doing research and other duties, he never showed it when I visited him. It was very strange to me. He never told me, "I am sorry but I am now too busy." Whenever I told him my ideas (for long hours), he always said, more than once, "That seems very interesting."[101]

While Nishina' role was central to the creation of Japan's first nationwide research network, Yukawa's meson theory contributed to its expansion. Since the early 1930s, theoretical physics had attracted talented young physicists such as Sakata, Kobayasi, Taketani and Tamaki, and they concentrated on the meson theory and related subjects. By working together and sharing their ethos with young professors like Yukawa and Tomonaga, elementary particle physics spread rapidly from 1940, when the young researchers became professors and began to train the next generation. A dramatic increase in the number of particle physicists produced in Japan between 1935 and 1950 proves how successfully this research network actually worked. According to statistics generated by Kaneseki Yoshinori, although only 7 theoretical physicists graduated from Japanese universities before 1930 (including Yukawa and Tomonaga), this number rose to 18 between 1930 and 1939 (including Kobayasi, Sakata, Taketani, and Tamaki), rose again to 41 between 1940 and 1945, and finally climbed to 71 between 1946 and 1949.[102]

The following dialogue about the "meson club (meson kai)" demonstrates that the meson theory provided young Japanese physicists with a locus working together:

> **Laurie M. Brown:** How did this Meson Symposium begin?
>
> **Yoichi Fujimoto:** Sakata writes that in November of 1937 Professor Nishina invited the theoretical group of Osaka and had a discussion meeting on mesons.

After that they had an informal discussion meeting regularly after the semi-annual meeting of Riken (IPCR) or annual meeting of the Physico-Mathematical Society. This was called *Meson-Kai, kai* meaning "club," "party," "discussion," or something like that.

Michiji Konuma: The meeting held in September 1943 was well organized and it was rather official; it was called *Chukanshi Toronkai* (Meson Symposium). In the next year, another meeting was held under the leadership of Satoshi Watanabe at the University of Tokyo. This was the last, because afterward the bombing got to be too bad.

Brown: Before we proceed, does everyone agree that the Meson Club began meeting in 1938 or 1939 and that all references to Meson Club, Meson Symposium, etc., refer to the same group? There seem to be printed records of only two meetings.

Konuma: No, only one. Proceedings of the Meson Symposium in September 1943.

Fujimoto: There were a number of meetings, less formal, to bring together people from the Osaka-Kyoto area.

Konuma: . . . According to an article of Taketani, when he moved from Kyoto to Tokyo in 1941, Nishina, Tomonaga, and Taketani talked about the plan to have informal meetings on the meson after the general twice-yearly meetings of Riken, for which people came from all over Japan. It was called *Meson-Kai* (Meson Club). At least several such meetings were held. Eventually they wanted to have a large two-day meeting so this one was held and called *Chukanshi-Toronkai* (Meson Symposium). This was the last large meeting before the end of the war, but there was another smaller meeting at the University of Tokyo later . . . .

Fujimoto: In 1942 there was a report on the two-meson theory.

Konuma: That was in the semi-annual meeting of Riken.

Brown: It is clear now. There was only one large two day meeting: that was in September 1943 and there the two-meson theories were discussed.

Konuma: Yes, but the two-meson theories were reported also in the regular meeting in June of 1942.

Fujimoto: In the meeting in September 1943, also the strong coupling theory was discussed by Tomonaga.[103]

Two points are apparent from this dialogue. The first is that Nishina was the patron of these meetings. Without his generous support, these meetings could not have continued over such a long period. Second, the meson theory enabled young Japanese theoreticians to overcome regional or factional divisions. It brought them together as colleagues and allowed them to identify themselves proudly as a distinctive group. The meson theory truly belonged to these young Japanese theoreticians, not to the physicists of Nishina's older generation.

Interestingly, Tokyo Imperial University played no distinguishing role in this very successful research network. In Tokyo, Riken with Nishina and Tomonaga formed the network's center, not the Department of Physics at Tokyo Imperial University. Former students who entered Tokyo Imperial University between 1939 and 1945 characterized Japan's most prestigious institute as too rigid to respond to the rapidly changing intellectual milieu outside. T. Takabayasi spent the 3 years between 1939 and 1941 in the department of physics at Tokyo Imperial University. He remembered the outdated research there:

> The line of studies which were particularly conspicuous were (a) instrumental physics and applied physics, along the tradition represented by Prof. Emeritus Seiji Nakamura, (b) "mathematical physics" under the patronage of Prof. Kan-ichi Terazawa, and (c) some "Terada physics." These do not appear to belong to main course of contemporary physics .... [T]he Department seems to have had an awkward personnel policy and maintained constitutional conservatism, making it difficult to open new domains of research.[104]

Nambu, who entered the University's department of physics in 1939, had a similar view:

> The University of Tokyo group was, up until my time, out of the mainstream of particle physics. The mainstream resided in Kyoto, Riken and Tokyo University of Education, the bases of Yukawa, Nishina and Tomonaga ....
>
> When I was a student in the prewar days, some of us students wanted to study particle physics, but there was no professor in that field. The only one who was really related to it was Professor K. Ochiai, but we were all discouraged from doing such study because of his remark that only geniuses can get into that field. We decided to organize ourselves and study and read papers ....
>
> In the prewar days, because we didn't have any active particle physics research inside the University of Tokyo, a few of us including Hayashi, went to listen to the Tomonaga-Nishina seminar at Riken. I learned quite a bit about cosmic ray physics. Tomonaga and Nishina participated in the same seminar. They would exchange ideas, and Tomonaga would discuss any letters he received at the time. For example, there were letters from Sakata in Nagoya. Of the two letters I remember, one was about π^0 decay. The question, I believe, was whether the pion came in an isospin triplet or just as charged pions. If you believe in isospin then you must assume the π^0. The question is then how to detect the π^0. Sakata pointed out that the decay mode, 2γ or 3γ, depends on the spin of the pion. If π is a vector meson, it cannot decay into 2γ, whereas if it is pseudoscalar it can. The next letter was just before I was drafted in the fall of 1942. Tomonaga reported on Sakata's letter on the two-meson theory in the seminar.[105]

In private interviews in 1999, Nambu repeated similar stories.[106] He remembered that he spent most of his time at the university in the library reading articles that had arrived from Europe and the United States. The most impressive professor in the physics department at that time was Kotani, a specialist in solid-state physics, who spoke softly and worked very hard. Nambu remembered that there always were mountains of papers filled with calculations in Kotani's office. Nambu also recalled that physics professors at Tokyo Imperial University regarded Yukawa's meson theory as the "pet theory of Kyodai (Kyoto Imperial Univeristy) physicists." For this reason, Nambu picked up what little he knew about meson theory from a semi-popular Japanese-language scientific journal. It was only after he and his friends attended Nishina–Tomonaga's seminar at Riken that he learned more about the meson theory. The seminar at Riken consisted of "about 20 attendees" and was always characterized by "very lively discussions on recent discoveries." Tomonaga was "the most frequent commentator as well as most severe critic."

Toichiro Kinoshita's personal recollections show that the conservatism of the physics department at Tokyo University persisted into the 1940s.

> I entered the University of Tokyo in September 1944, as a beginning student in the department of physics .... The most memorable thing that happened in this period [was] a series of three or four lectures given by Professor Yukawa, an introduction to quantum electrodynamics, in the spring of 1945. He was a professor at Kyoto University but came to Tokyo to educate us at a considerable risk to his own safety. I suppose one reason for such an arrangement was the fact that the physics department of Tokyo University then had nobody on its faculty who was actively engaged in the research of particle physics.[107]

The brilliant success of Yukawa and Tomonaga seemed to have no impact on the conservative attitude of physicists at Tokyo Imperial University.

Ironically, however, the rigid traditionalism of Tokyo Imperial University contributed greatly to the successful development of a nationwide research network that overcame Japan's regional and academic factionalism. During this process, Nishina, the network's true patron, became the beloved *sensei* of all Japanese theoreticians. However, his contribution to creating pathways of communication throughout the Japanese physics community did not end with this remarkable achievement. Nishina fostered two other research groups, one focused on cosmic ray research and the other on nuclear physics. As discussed in the next two chapters, their influence would prove equally important to the development of physics in Japan.

NOTES

1 "Nuclear Research at Riken: Dialogue with the late Sin-itiro Tomonaga," in Laurie M. Brown, Michiji Konuma, and Ziro Maki (eds.), *Particle Physics in Japan, 1930–1950,* Vol. II (Kyoto: Research Institute for Fundamental Physics, Kyoto University, 1980), pp. 1–25 on p. 7.

2 H. Yukawa, "Nishina-sensei, Tomonaga-san and I [in Japanese]," *Nishina Memorial Lectures*, Vol. 7 (Tokyo: Nishina Memorial Foundations 1966), p. 12.

3 S. Tomonaga, "Paradise for Scientists [in Japanese]," in *Collected Essays of Tomonaga Sin-itiro*, (Tokyo: MYU, 1995), Vol. 1, pp. 224–235 on pp. 224–225.

4 Ibid., p. 224.

5 H. Tamaki, "Memories of the Theory Group, Nishina Laboratory," in Matsui and Ezawa (eds.), *Sin-itiro Tomonaga*, (Tokyo: MYU, 1995), pp. 127–132 on p. 128.

6 Minoru Kobayasi, "Reminiscences of Our Days at the Institute of Physical and Chemical Research," in Matsui and Ezawa (eds.), *Sin-itiro Tomonaga*, (Tokyo: MYU, 1995), pp. 133–139 on pp. 133–134.
7 Nobuyuki Fukuda, Yoneji Miyamoto and Tatsuoki Miyazima, "To Sin-itiro Tomonaga on his Sixtieth Birthday," *Supplement of the Progress of Theoretical Physics: Dedicated to Professor Sin-itiro Tomonaga on the Occasion of His Sixtieth Birthday*, *37 & 38* (1966), iii–viii on iii.
8 Takeo Hori, "Bohr's Group and Copenhagen Spirit," in *Nishina Yoshio*, pp. 46–54.
9 John L. Heilbron, "The Earliest Missionaries of the Copenhagen Spirit," in Edna Ullmann-Margalit (ed.), *Science in Reflection* (Dordrecht: Kluwer Academic Publishers, 1988), pp. 201–233.
10 Tomonaga, "Last Lecture," in *The Story of Spin*, pp. 226–227.
11 See Akira Shimizu, "Dr. Tomonaga and Tozan-so," in Matsui and Ezawa (eds.), *Sin-itiro Tomonaga*, (Tokyo: MYU, 1995), pp. 119–120.
12 Y. Nishina and S. Tomonaga, "On the Creation of Positive and Negative Electrons," *Proc. Phys. Math. Soc.*, *15* (1933), 248–249. The first quotation is from "Nishina Laboratory" in "Summary of the Past activities of the Institute," *SP*, *34* (1938), 1842–1849 on 1842.
13 S. Tomonaga, "A Farewell to Dr. Sakata," in Matsui and Ezawa (eds.), *Sin-itiro Tomonaga*, (Tokyo: MYU, 1995), pp. 115–118 on pp. 115–116.
14 Y. Nishina and S. Tomonaga, "Problems Related to Positive Electron [in Japanese]," *Kagaku*, *3* (1933), 390–393.
15 G. Beck, "Hat das negative Energiespektrum einen Einfluß auf Kernphänomene?" *Zeitschrift für Physik*, *83* (1933), 498–511; J. R. Oppenheimer and M. S. Plesset, "On the Production of the Positive Electron," *Physical Review*, *44* (1933), 53–55. Beck's paper was received in May 2, and Oppenheimer and Plesset's letter was dated in June 9.
16 W. Heitler and F. Sauter, "Stopping of Fast Particles with Emission of Radiation and the Birth of Positive Electrons," *Nature*, *132* (1933), 892.
17 Y. Nishina, S. Tomonaga and Shoichi Sakata, "On the Photo-Electric Creation of Positive and Negative Electrons," *Supplement to SP*, *17* (1934), 1–5.
18 Hans Bethe and W. Heitler, "On the Stopping of Fast Particles and on the Creation of Positive Electrons," *Proceedings of the Royal Society*, *146* (1934), 83–112. The quotation comes from "Nishina Laboratory," *SP*, *34* (1938), 1842.
19 Y. Nishina and S. Tomonaga, "On the Negative-Energy Electrons," *Japanese Journal of Physics*, *9* (1933–1934), 35–40.
20 Ibid., 35.
21 Ibid., 38–39.
22 Y. Nishina, S. Tomonaga, and H. Tamaki, "On the Annihilation of Electrons and Positrons," *Supplement to SP*, *18* (1934), 7–12.
23 Y. Nishina, S. Tomonaga, and M. Kobayashi, "On the Creation of Positive and Negative Electrons by Heavy Charged Particles," *SP*, *27* (1935), 137–178.
24 Ibid., 138–139.
25 L. Landau and E. Lifshitz, "On the Production of Electrons and Positrons by a Collision of Two Particles," *Physikalische Zeitschrift der Sowjetunion*, *6* (1934), 244–257; E. J. Williams, "Production of Electron–Positron Pairs," *Nature*, *135* (1935), 66; W. Heitler and L. Nordheim, "Sur la production des paires par des chocs de particules lourdes," *Journal de Physique*, *5* (1934), 449–454; J. R. Oppenheimer, "Note on the Production of Pairs by Charged Particles," *Physical Review*, *47* (1935), 146–147; L. Nordheim, "Sur la production des paires par des chocs de particules," *Journal de Physique*, *6* (1935), 135–136.

26 Y. Nishina, S. Tomonaga, and H. Tamaki, "A Note on the Interaction of the Neutron and the Proton," *SP*, *30* (1936), 61–69.

27 Ibid., 62.

28 Hans A. Bethe, "Theory of Disintegration of Nuclei by Neutrons," *Physical Review*, *47* (1935), 747–759. See also H. Bethe and R. Peierls, "The Scattering of Neutrons by Protons," *Proceedings of the Royal Society*, *149* (1935), 176–183.

29 S. Tomonaga, "Last Lecture," in *The Story of Spin*, p. 228.

30 Y. Nishina to P. A. M. Dirac (date not known, probably in the spring of 1931), *P. A. M. Dirac–Y. Nishina Correspondence, 1928–1948, No. 33* (Tokyo: Nishina Memorial Foundation, 1990), p. 11.

31 P. A. M. Dirac to Y. Nishina (April 22, 1931), ibid., p. 11.

32 Y. Nishina to P. A. M. Dirac (June 15, 1931), ibid., p. 13; Y. Nishina to P. A. M. Dirac (August 10, 1931), ibid., p. 15.

33 P. A. M. Dirac to Y. Nishina (July 10, 1931), ibid., p. 14.

34 P. A. M. Dirac to Y. Nishina (October 6, 1934), ibid., p. 17.

35 Hidehiko Tamaki, "Memories of the Theory Group, Nishina Laboratory," in Matsui and Ezawa (eds.), *Sin-itiro Tomonaga*, pp. 127–132 on p. 129.

36 Ibid., p. 131.

37 Y. Nishina to P. A. M. Dirac (February 11, 1936), *P. A. M. Dirac–Y. Nishina Correspondence, 1928–1948*, pp. 24–25 on p. 25.

38 Ibid., p. 24.

39 Y. Nishina to P. A. M. Dirac (December 16, 1948), ibid., No. 33, p. 31.

40 K. Umeda, S. Tomonaga and Y. Ono, "Eine Bemerkung über die gegenseitigen potentiellen Energien zwischen zwei Deuteronen," *SP*, *32* (1937), 87–96.

41 S. Tomonaga and K. Umeda, "Eine Bemerkung zum Austauschintegral," *SP*, *32* (1937), 97–102.

42 S. Tomonaga, "Bemerkungen über die kinetische Kernenergie im Hartree–Fock–Modell," *SP*, *32* (1937), 229–232.

43 S. Tomonaga and H. Tamaki, "On the Collision of a High Energy Neutrino with a Neutron," *SP*, *33* (1937), 288–298.

44 Ibid., 296–297.

45 S. Tomonaga and M. Kobayasi, "Scattering and Splitting of Photons on the Non-Linear Field Theory of Born and Infeld," *SP*, *34* (1938), 1643–1649. They read the summary of the paper in November 1935 at the semiannual meeting of Riken.

46 K. Umeda and Y. Ono, "Über das Flüggesche Atomkernmodell," *Bulletin IPCR*, *15* (1936), 674–680.

47 K. Umeda and Y. Ono, "Über das dynamische Flüssigkeitsmodell des Atomkerns," *SP*, *32* (1937), 120–128.

48 K. Umeda, "Zum Amplitudenfaktor in der Fermischen β-Zerfallsmatrix," *SP*, *34* (1938), 137–143; K. Umeda, "Die Termabstände der Atomkerne nach dem Oszillatormodell," *SP*, *34* (1938), 197–204; K. Umeda, "Zur Beziehung zwischen Partitio Numerorum und Kernanregung," *SP*, *34* (1938), 629–636; K. Umeda, "Über die Debyetemperatur des Flüssigkeitströpfchenmodells für den Atomkern," *SP*, *35* (1939), 8–15; K. Umeda, "Über den Beitrag der Schwankungsbindung zur Kernanregungsenergie," *SP*, *36* (1939), 57–71.

49 K. Ariyama, "Über die Zustände der Elektronen der zweiwertigen Metalle," *SP*, *34* (1938), 344–356.

50 M. Kobayasi, "Thomas–Fermi Method and Bohr's Nuclear Model," *Bulletin IPCR*, *16* (1937), 219–227.

51 M. Kobayasi and S. Ozaki, "On the Energy Loss of Fast Charged Particles by Pair Creation," *SP*, *34* (1938), 321–331.
52 Ibid., 331.
53 S. Tomonaga, "Nuclear Research at Riken," in Laurie M. Brown, Rokuo Kawabe, Michiiji Konuma, and Ziro Maki (eds.), *Elementary Particle Theory in Japan, 1935–1960* (Kyoto: Research Institute for Fundamental Physics, University of Kyoto, 1988), pp. 1–25 on p. 3.
54 S. Tomonaga, "Innere Reibung und Wärmeleitfähigkeit der Kernmaterie," *Zeitschrift für Physik*, *110* (1938), 573–604.
55 Hans Euler and Werner Heisenberg, "Theoretische Gesichtspunkte zur Deutung der kosmischen Strahlung," *Ergebnisse der Exacten Naturwissenschaften*, *17* (1938), 1–69.
56 "Nuclear Research at Riken," in Brown, Konuma, and Maki (eds.), *Particle Physics in Japan, 1930–1950*, Vol. II, p. 4.
57 M. Kobayasi, "Reminiscences of Our Days at the Institute of Physical and Chemical Research," in *Sin-itiro Tomonaga*, pp. 133–139 on p. 138.
58 H. Yukawa, "On the Interaction of Elementary Particles. I," *Proc. Phys. Math. Soc.*, *17* (1935), 48–57 on 57.
59 For the history of the development of meson theory, see Viśvapriya Mukherji, "A History of the Meson Theory of Nuclear Forces from 1935 to 1952," *Archive for History of Exact Science*, *13* (1974), 27–102; Laurie M. Brown and Lillian Hoddeson (eds.), *The Birth of Particle Physics* (Cambridge: Cambridge University Press, 1983); Olivier Darrigol, "The Quantum Electrodynamical Analogy in Early Nuclear Theory or the Roots of Yukawa's Theory," *Revue d'Histoire des Sciences et de leurs Applications*, *41* (1988), 225–297; Laurie M. Brown, "Yukawa in the 1930s: A Gentle Revolutionary," *Historia Scientiarum*, *36* (1989), 1–21; Helmut Rechenberg and Laurie M. Brown, "Yukawa's Heavy Quantum and the Mesotron (1935–1937)," *Centaurus*, *33* (1990), 214–252; Laurie M. Brown and Helmut Rechenberg, "The Development of the Vector Meson Theory in Britain and Japan (1937–38)," *British Journal for the History of Science*, *24* (1991), 405–433.
60 Laurie M. Brown, "Chapter 5. Nuclear Forces, Mesons, and Isospin Symmetry," in Laurie M. Brown, et al. (eds.), *Twentieth Century Physics*, (Princeton: Princeton University Press, 1999) Vol. I, pp. 357–419 on pp. 389–409.
61 See Olivier Darrigol, "Elements of a Scientific Biography of Tomonaga Sin-itiro," *Historia Scientiarum*, *35* (1989), 1–29.
62 S. Tomonaga and G. Araki, "Effect of the Nuclear Coulomb Field on the Capture of Slow Mesons," *Physical Review*, *58* (1940), 90.
63 Ibid.
64 S. Tomonaga, "Über den Zusammenstoß des Mesotrons mit Elektronen," *SP*, *37* (1940), 399–413.
65 S. Tomonaga, "Zur Theorie des Mesotrons. I," *SP*, *39* (1941), 247–266; T. Miyazima and S. Tomonaga, "Zur Theorie des Mesotrons. II," *SP*, *40* (1942), 21–67.
66 S. Tomonaga (1941), "Zur Theorie I," 247.
67 S. Tomonaga, "Bemerkung über die Streuung der Mesotronen am Kernteilchen," *SP*, *40* (1942), 73–86.
68 T. Miyazima and S. Tomonaga, "On the Mesotron Theory of the Nuclear Forces," *SP*, *40* (1943), 274–310.
69 Ibid., 275.
70 Ibid., 276.
71 Silvan S. Schweber, *QED and the Men Who Made It: Dyson, Feynman, Schwinger, and Tomonaga* (Princeton: Princeton University Press, 1994), p. 260.

72 S. Tomonaga, "On the Interaction between Mesons and Nucleons I," *Bulletin IPCR*, *22* (1943), 38–40; S. Tomonaga, "On the Interaction between Mesons and Nucleons II," *Bulletin IPCR*, *22* (1943), 43–44.

73 M. Kobayasi and R. Utiyama, "Note on the Pair Creation of Mesons by γ-Rays and the 'Bremsstrahlung' of Mesons in the Nuclear Field," *SP*, *37* (1940), 221–225.

74 G. Araki, "Production and Capture of Pseudoscalar Mesons," *SP*, *39* (1941), 14–27; G. Araki, "Nuclear Scattering of Pseudoscalar Mesons," *SP*, *40* (1943), 311–330.

75 T. Miyazima, "On the Mesotron of Spin One-Half," *SP*, *39* (1941), 28–40.

76 S. Ozaki, "On the Scattering of Mesons by the Electrostatic Field with Spherical Well Potential," *SP*, *39* (1941), 223–246 on 224.

77 H. Tamaki, "Pri la al duaŝtufa undoro operaciendaj operatoroj," *SP*, *40* (1942), 11–20.

78 K. Ariyama, "Zur Theorie der Supraleitung," *SP*, *39* (1941), 148–156; H. Tamaki, "Pri la ondoekvacio de korpuskloj kun entjerplusduona ŝpino," *SP*, *40* (1942), 1–10.

79 S. Tomonaga, "On a Relativistic Reformulation of Quantum Field Theory," *Bulletin IPCR*, *22* (1943), 545–557. See Schweber, *QED and the Men Who Made It*, pp. 260–272.

80 Laurie M. Brown and Y. Nambu, "Physicists in Wartime Japan," *Scientific American* (December, 1998), 96–103 on 99–100.

81 Tanikawa, Okayama, and Hai (an unidentified Korean) were also members of Yukawa's Osaka group.

82 S. Kikuchi, H. Aoki, and K. Husimi, "The Emission of the Electron from the Substances Traversed by Fast Neutrons," *Proc. Phys. Math. Soc.*, *18* (1936), 725–744 on 744; H. Aoki, "Excitation of Gamma-Rays by Fast Neutrons," *Proc. Phys. Math. Soc.*, *19* (1937), 369–385 on 376.

83 H. Yukawa and S. Sakata, "On the Theory of Collision of Neutrons with Deuterons," *Proc. Phys. Math. Soc.*, *19* (1937), 542–551 on 547; H. Yukawa and S. Sakata, "On the Interaction of Elementary Particles II," *Proc. Phys. Math. Soc.*, *19* (1937), 1084–1093 on 1090.

84 "Nuclear Research at Osaka University," in Laurie M. Brown, et al. (eds.), *Particle Physics in Japan, 1930–1950*, Vol. II, pp. 39–47 on pp. 42–43. Italics are added.

85 Yukawa, *Tabibito*, p. 203.

86 Sin-itiro Tomonaga, "Reminiscences," *Supplement of the Progress of Theoretical Physics: Dedicated to Professor Minoru Kobayasi on the Occasion of His Sixtieth Birthday*, (1968), 3–5 on 3.

87 For this episode, see Mitsuo Taketani, "Methodological Approaches in the Development of the Meson Theory of Yukawa in Japan," *Supplement of the Progress of Theoretical Physics: Philosophical and Methodological Problems in Physics*, *50* (1971), 12–26, especially on 19–23.

88 H. Yukawa, "On the Interaction of Elementary Particles I," 52.

89 M. Kobayasi and T. Okayama, "On the Creation and Annihilation of Heavy Quanta in Matter," *Proc. Phys. Math. Soc.*, *21* (1939), 1–13 on 13.

90 M. Taketani and S. Sakata, "On the Wave Equation of Meson," *Proc. Phys. Math. Soc.*, *22* (1940), 757–770 on 770.

91 M. Kobayasi and R. Utiyama, "On the Interaction of Mesons with Radiation Fields," *Proc. Phys. Math. Soc.*, *22* (1940), 882–898 on 898.

92 Yukawa interview with John A. Wheeler (July 10, 1962), AIP OH 575, p. 5.

93 Ibid.

94 H. Yukawa, "Nishina-sensei, Tomonaga-san and I," p. 12.

95 Seitaro Nakamura, *Yukawa Hideki and Tomonaga Sin-itiro* [in Japanese] (Tokyo: Yomiuri Shinbun, 1992), pp. 61 and 76.

96 Yoichiro Nambu, "Directions of Particle Physics," *Supplement of the Progress of Theoretical Physics: Proceedings of the Kyoto International Symposium: The Jubilee of the Meson Theory, Kyoto, August 15–17, 1985, 85* (1985), 104–110 on 104.

97 H. Yukawa, "On the Interaction of Elementary Particles I," 57.

98 H. Yukawa and C. Kikuchi, "The Birth of the Meson Theory," *American Journal of Physics, 18* (1950), 154–156 on 155. See also Yukawa interview with John A. Wheeler, p. 2. For the mentioned 1933 presentation ("A Comment on the Problem of Electrons in the Nucleus"), see Laurie M. Brown, et al. (eds.), *Elementary Particle Theory in Japan, 1935–1960*, p. 157.

99 Yukawa, *Tabibito*, p. 203.

100 "Particle Physics in Japan in the 1940s," in Laurie M. Brown, Michiji Konuma and Ziro Maki. (eds.), *Particle Physics in Japan, 1930–1950*, Vol. 1, (Kyoto: Research Institute for Fundamental Physics, Kyoto University, 1980) pp. 43–71 on p. 48. For this 1934 presentation ("On the Probability Amplitude in Relativistic Quantum Mechanics"), see Laurie M. Brown, et al. (eds.), *Elementary Particle Theory in Japan, 1935–1960*, pp. 169–173.

101 H. Yukawa, "Nishina-sensei, Tomonaga-san and I," pp. 8 and 10.

102 K. Yoshinori, "The Elementary Particle Theory Group," in Nakayama, et al. (eds.), *Science and Society in Modern Japan*, pp. 221–252 on pp. 222–229.

103 "Particle Physics in Japan in the 1940s," in Laurie M. Brown, et al. (eds.), *Particle Physics in Japan, 1930–1950*, Vol. 1, pp. 43–67 on pp. 56–58.

104 T. Takabayasi, "Physics in Tokyo University just before the Pacific War," in *Particle Physics in Japan, 1930–1950*, Vol. 1, p. 62. Terada physics refers to a phenomenological type of physics, inspired by everyday observations and reflecting, like the Japanese style of mini-poetry, *Haiku*, a typically Japanese mentality. This study was motivated partly by a desire for independence from Western physics and contained amusing, imaginative ideas that undoubtedly could have led to deeper theories.

105 Y. Nambu, "Summary of Personal Recollection of the Tokyo Group," in Laurie M. Brown, et al. (eds.), *Elementary Particle Theory in Japan, 1935–1960*, pp. 3–6 on pp. 3–4.

106 Y. Nambu interview with Dong-Won Kim (held in the University of Chicago, March 19 and May 13–14, 1999).

107 T. Kinoshita, "Personal Recollections, 1944–1952," in Laurie Brown, et al. (eds.), *Elementary Particle Theory in Japan, 1935–1960*, (Kyoto: Research Institute for Fundamental Physics, University of Kyoto, 1988), pp. 7–10 on p. 7.

5 Cosmic Ray Research: "This Is Interesting. Let Us Try It."

The history of cosmic rays is a history of failures and difficulties as well as successes. This is true of all fields of research but seems to be especially true in the case of cosmic rays. This arises partly from the difficulty of detecting the radiation in the first place but also from the wholly unexpected nature of the many phenomena encountered. Almost everyone in the game made mistakes. These involved errors not only in the taking of the data because of unknown influences at work but also in the interpretation of even the good data.

H. Victor Neher[1]

Masa Takeuchi: Nishina-san was very interested in building apparatus. On the contrary, he was not very eager to take idea (laughing).

Hidehiko Tamaki: He was always more interested in the bigness of the apparatus, rather than in the design of the apparatus itself.

Sin-itiro Tomonaga: Nishina-san realized the necessity of large experiment devices after looking at the example of Blackett. Thus he was eager to make a larger magnet and arrive at the level of Blackett, and then to go ahead.

Dialogue with Sin-itiro Tomonaga[2]

Silvan S. Schweber: At Los Alamos there was a sense of community and cooperation in which people talked freely with each other. Did you have a similar sense of community in Tokyo?

Satio Hayakawa: Such a community had existed at Riken [in the Nishina Laboratory] even before the war. Theoreticians and experimentalists freely discussed their work and their findings with each other.

Dialogue with the Tokyo Group[3]

5.1 COSMIC RAY RESEARCH IN THE EARLY TWENTIETH CENTURY

In 1930, cosmic ray physics was still a fresh new field that had been ushered in at the turn of century. In 1901, C. T. R. Wilson, Hans Geitel, and Julius Elster independently discovered in the atmosphere the presence of penetrating radiation

FIGURE 5.1 Victor F. Hess, the pioneer of cosmic ray research. (Courtesy of AIP Emilio Segrè Visual Archives.)

whose source was unknown. Some suggested an extraterrestrial source. By 1910, the balloon experiment was initiated to determine whether the intensity of the radiation depend on the altitude.

The first extensive and systematic research program to find the origin of this penetrating radiation was carried out by Victor F. Hess at the Institute for Radium Research of the Austrian Academy of Sciences (Figure 5.1).[4] On August 7, 1912, Hess made a balloon ascent to a height of 5000 m, bringing with him an airtight ionization measuring apparatus "fitted with a sensitive electromagnetic system which was not influenced by the large fluctuations of temperature occurring in the flight."[5] His apparatus clearly indicated a rise in intensity of the mysterious radiation at high altitude, proving that this radiation, which he and other German scientists called "Höhenstrahlen" (radiation from above), had an extraterrestrial source. Hess was hailed as the discoverer of these rays, and in 1936 he shared the Nobel Prize in physics with Carl D. Anderson. In his Nobel Lecture, Hess briefly explained his experiment:

> When, in 1912, I was able to demonstrate by means of a series of balloon ascents, that the ionization in a hermetically sealed vessel was reduced with increasing height from the earth (reduction in the effect of radioactive substances in the earth), but that it noticeably increased from 1000 m onwards, and at 5 km height reached several times the observed value at earth level, I concluded that *this ionization might be attributed to the penetration of the earth's atmosphere from outer space by hitherto unknown*

FIGURE 5.2 Robert A. Millikan made a great contribution to the development of cosmic ray research after the end of World War I. (Courtesy of the Archives, California Institute of Technology.)

> *radiation of exceptionally high penetrating capacity*, which was still able to ionize the air at the earth's surface noticeably. Already at that time I sought to clarify the origin of this radiation, for which purpose I undertook a balloon ascent at the time of a nearly complete solar eclipse on the 12th April 1912, and took measurements at heights of two to three kilometres. As I was able to observe no reduction in ionization during the eclipse I decided that, essentially, the sun could not be the source of cosmic rays, at least as far as undeflected rays were concerned. Many esteemed physicists in Europe and America have tried since then to solve the problems of the origin of cosmic rays.[6]

In 1913–1914, Werner Kolhörster of Berlin's Physikalisch-Technische Reichsanstalt ascended to 9000 m to perform a similar experiment that confirmed that ionization strength became much more intense at high altitude. He also discovered that the rays were far more penetrating than any known γ-rays. Most scientists then suggested that the rays were especially strong γ-rays (ultra γ-rays). Research concerning these mysterious rays was brought to an abrupt halt by the outbreak of World War I in the summer of 1914.

When research on cosmic rays resumed after the end of the war, an influential American physicist, Robert A. Millikan, entered the race to discover their source (Figure 5.2). Millikan, the 1923 Nobel laureate in physics, and his colleagues at the newly founded California Institute of Technology carried out a series of experiments in balloons and airplanes as well as in "snow-fed lakes at high altitude."[7] He not only confirmed that the rays definitely come in from above and that their origin is entirely outside the layer of atmosphere but also gave the rays the name that stuck: cosmic rays. Millikan suggested that primary cosmic rays were photons released when extraterrestrial photons and electrons combined in space to form elements like helium, oxygen, and carbon. The cosmic rays were none other than "the birth cries" of infant atoms, an exciting idea that captured the popular imagination.[8] Although Millikan's theory did not survive long, cosmic ray research continued to flourish at Caltech. The Nobel laureate, Carl D. Anderson, while a research fellow at Caltech, discovered the positron in 1932 and later, with his first graduate student, Seth Neddermeyer, identified the mesotron in 1936.

During the latter half of the 1920s, many important discoveries made in Europe and the United States cast serious doubt on the photon (γ-ray) theory of cosmic rays. In 1927, J. Clay of Amsterdam published data that he had gathered on a voyage from Amsterdam to Java. He observed that cosmic rays had less intensity near the equator, where the horizontal component of the geomagnetic field is stronger than at higher latitudes.[9] This latitude effect indicated that primary cosmic rays, before entering into the terrestrial atmosphere, might be charged particles that were affected by the geomagnetic field.

Also that year, Dmitry V. Skobeltzyn, a young researcher at Leningrad Physicotechnical Institute, succeeded in observing fast-moving cosmic ray particles in a Wilson cloud chamber placed in a strong magnetic field.[10] To determine whether Skobeltzyn's particles were primary cosmic rays or of secondary origin, Walther Bothe of Berlin and Kolhörster of Potsdam developed the new coincidence counting technique, consisting of two Geiger–Müller counters heavily shielded from earth-originated radiations and from each other by thicknesses of iron (5 cm), lead (6 cm), and gold (4.1 cm). This armor, however, did not decrease the rate of

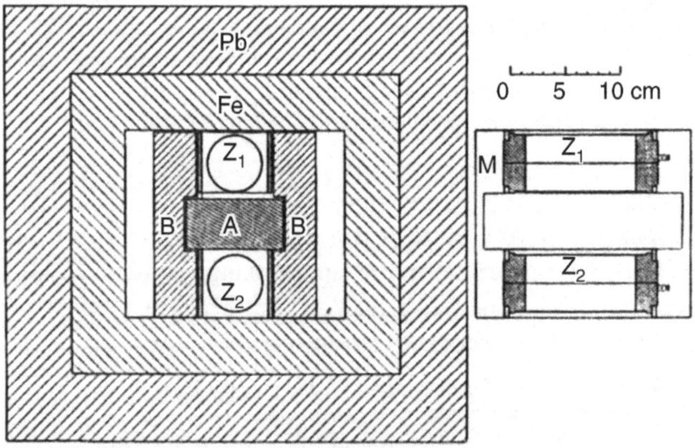

FIGURE 5.3 The coincident counter method for cosmic rays research was first introduced by Walter Bothe and Werner Kolhörster in 1929. (W. Bothe and W. Kolhörster, "Das Wesen der Höhenstrahlung," *Zeitschrift für Physik, 56* [1929], 751–777 on 754.) Their experiment indicated the corpuscular nature of cosmic ray primaries.

what they called "coincidental" registrations of particle impacts on their electroscopes. Based on the results of their experimentation, Bothe and Kolhörster in 1929 concluded that it was likely that cosmic rays were not waves but "a corpuscular radiation," (i.e. a stream of moving particles) of enormous velocity.[11] Bruno Rossi, then an assistant at the Physics Institute of the University of Florence, took their method a step further with his fast coincidence electron counter, which contained a row of three Geiger counters to record the simultaneous occurrence of three coincident electrical pulses. Rossi's results showed that "cosmic rays were capable of producing an enormously more abundant secondary radiation than any other known rays."[12]

By the early 1930s, cosmic ray research had become very interesting not only to experimentalists but also to top theoretical physicists, including Heisenberg. At two international conferences held in London in 1931 and 1934, cosmic rays were a central topic. Newly developed instruments, such as Rossi's multiple-coincidence circuit and P. M. S. Blackett and G. Occhialini's cloud chamber controlled by Geiger–Müller counters, revealed more of the puzzling nature of cosmic rays, including their soft and hard components, pair-creation, and the cosmic ray shower phenomenon. All of these discoveries required theoretical interpretations, and Anderson's 1932 discovery of the positron drew experimentalists and theoreticians together. As Satio Hayakawa observed, theoreticians "began to pay great attention" to cosmic ray physics "as a field open to the development of theory," and experimentalists began to recognize quantum mechanics as an important experimental tool.[13] For theoreticians, cosmic rays "were the natural experimental testing ground for fundamental theories," and they soon were supplying plausible explanations for cosmic ray phenomena discovered by experimentalists.[14] In 1936, for example, Heisenberg proposed his theory of explosive showers; that year Homi J. Bhabha and Walter H. Heitler, using Bethe–Heitler calculations, proposed the cascade theory of cosmic ray showers.

5.2 THE EFFECT OF THE 10TH SUBCOMMITTEE OF THE JAPAN SOCIETY FOR THE PROMOTION OF SCIENCE ON COSMIC RAY RESEARCH IN JAPAN

In Japan, cosmic ray research started with the opening of the Nishina Laboratory in the summer of 1931. Ryokichi Sagane and Masa Takeuchi were the first scientists recruited for this line of research. Sagane, a son of Nagaoka, who had been adopted by the Sagane family through marriage, was a graduate of the Department of Physics at Tokyo Imperial University. Takeuchi was a graduate of Tokyo Engineering High School, where he had studied applied chemistry. Together, in 1932, they built a small cloud chamber for cosmic ray research and several different types of Geiger–Müller counters, one of which was used to detect changes of cosmic ray intensities when a typhoon passed through the Tokyo area (Figure 5.4).[15] The capabilities of this apparatus were disappointing, as Takeuchi later recalled:

> The first cloud chamber we made was expanded by a horizontally moving piston, and was mounted in a magnetic field of 2000 Gauss, which was generated by a Helmholtz coil. At the end of 1932, we succeeded in taking photographs of cosmic ray tracks.

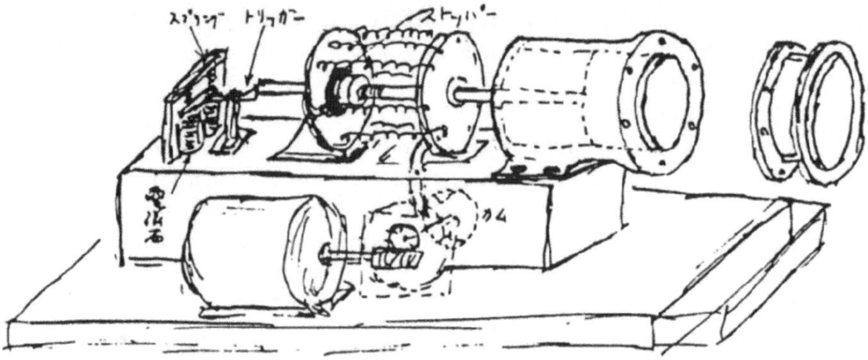

FIGURE 5.4 Early drawing of the cloud chamber that the cosmic ray research group in the Nishina Laboratory constructed in 1932. (Courtesy of the Institute of Physical and Chemical Research.)

> The counter controlled method for cloud chamber operation was invented by R. Sagane, and was applied to this experiment as mentioned in November 1932 at the meeting of I.P.C.R. [Riken] . . . .
>
> In March of 1933, Blackett published the result obtained by the counter controlled cloud chamber in a magnetic field (Blackett and Occhialini, 1933). Their experiment was similar to ours, but with a stronger magnetic field. Having read this paper, we found that the pair of circular tracks in our photograph of 1932 was nothing but a pair creation and that the picture with multiple tracks showed a shower phenomenon. However, we could not but conclude that our apparatus was not good enough to continue further experiments on energy spectrum, and that a larger cloud chamber in a stronger magnetic field was necessary for the future study of high energy cosmic ray particles. Thus we decided to suspend the cloud chamber experiment for a while.[16]

Nishina, Sagane, and Takeuchi reported on this experimentation at the semi-annual meeting of Riken in November of 1933 in a paper entitled "Research on Cosmic Rays using the Wilson Cloud Chamber (preliminary)."[17]

The desired turning point came with the establishment of the 10th subcommittee of the Japan Society for the Promotion of Science (Nihon Gakujutsu Shinkokai). The Japan Society was founded in December of 1932 with generous donations from the emperor, the government, and industry.[18] The Japan Society operated 12 committees (including three for humanities and social sciences), 100 subcommittees, and 30 special committees "to promote scientific research and its applications, to accelerate cultural, industrial and military development, and to contribute to the prosperity of Japan as well as the welfare of the world." It had great impact on scientific development in Japan during the 1930s. The Japan Society generously funded science and engineering projects, particularly practical projects in applied science, medicine, and engineering. During the 1930s, these projects included Nishina's projects on cosmic rays and nuclear physics as well as organized efforts to develop new airplane fuel and special steel, improve wireless communication, create a national standard of nutrition, research on tuberculosis and Japanese encephalitis, and so on.

Each of these projects operated with budgets several times larger than the Ministry of Education's entire annual budget for scientific research.[19] Through them, the Japan Society encouraged cooperative team research on a national scale, breaking the Japanese tradition of individual efforts in university laboratories. Each major project was regulated by a subcommittee or special committee especially created for that project.

Of the Japan Society's many subcommittees and special committees, the 10th subcommittee may have been the most important for the development of science in Japan, chiefly because of the great consequence of its research subjects: cosmic rays and the nuclear physics.

> The aim of this committee is to investigate cosmic rays and the atomic nucleus together since they are closely related to each other. For the investigation of cosmic rays, the research aims to elucidate the nature and the generating mechanism of cosmic rays, to inquire into their interaction with matter, to investigate effects on geophysical phenomena and to study biological effects. First of all, the committee will measure the intensities of cosmic rays in different parts of Japan and the world. Then, the relationship between these data and geophysical (meteorology, geomagnetism, geophysics, K–H electrosphere) and astronomical phenomena will be investigated. Finally, it will cooperate with other countries to gather information about the intensities of cosmic rays in the surface as well as the atmosphere of the entire earth.
>
> The sub-section for the atomic nucleus is to study the structure of the atomic nucleus, the artificial transformation of elements, artificial radioactivity and the like, and to pursue the application of results to medicine, technology, chemistry, and biology.[20]

Established on January 29, 1934, the membership of the 10th subcommittee consisted of very important scientific figures: Takematsu Okada, director of the Central Meteorological Observatory and first chairman of the 10th subcommittee; Torahiko Terada and Mishio Ishimoto, both from Tokyo Imperial University; and Masao Kinoshita and Nishina, both from Riken. Masaharu Nishikawa of Riken joined the committee in May 1934. Three years later, in March of 1937, after nuclear research had emerged as the more important of the subcommittee's two subjects, four new members were added: Nagaoka of Riken, who had become the new chairman, Bunsaku Arakatsu of Kyoto Imperial University, Seishi Kikuchi of Osaka Imperial University, and Yoshikatsu Sugiura of Riken. Supported by these influential scientists, Nishina's cosmic ray group almost monopolized Japanese research in that field starting in the mid-1930s. A competing small cosmic ray group was started in 1934 at Osaka Imperial University by Seishi Kikuchi, with Yuzuru Watase and Ichimiya.[21] This Osaka group developed a coincidence method and found evidence for the nucleon cascade, but, after Kikuchi turned his attention to nuclear physics, this group lost its importance.

Among the members of the 10th subcommittee, Okada deserves special mention. Japanese historian of science, Tetu Hirosige, argued that Okada made the greatest contribution to the success of the 10th subcommittee, at least in its early stages.[22] An influential professor at Tokyo Imperial University, Okada sat on various important governmental and academic bodies, including the Imperial Academy, the National Research Council, and the Board of Examiners for Air Transport.

He also was the director of the Central Meteorological Observatory and the president of the Meteorological Society of Japan. He believed that cosmic ray research might help modernize meteorology and the weather service. As the first chairman of the 10th subcommittee, he actively garnered support for cosmic ray research. Consequently, many of the cosmic ray group's research was related to meteorology, a circumstance that later provided the Nishina Laboratory with convenient excuses to continue cosmic ray research during the last days of World War II.

Nishina was appointed as both chief cosmic ray researcher and secretary to the 10th subcommittee. As such, he enjoyed the strong support of other members of the subcommittee. The 10th subcommittee provided Nishina with two important assets: funding and a network. More funding meant more and better experimental instruments, and a wider network brought talented researchers to the Nishina Laboratory from all over Japan. Beginning in 1936, when the assets — funding and network — of the 10th subcommittee were fully realized, Nishina's cosmic ray group began to publish important results.

5.3 THE COSMIC RAY RESEARCH GROUP

The cosmic ray research group in the Nishina Laboratory shared several characteristics with the Laboratory's theory research group. Both groups started at almost the same time in the 1931–1932 term. Like the theory group, the cosmic ray group was comprised by researchers with different educational backgrounds: Masa Takeuchi studied applied chemistry at Tokyo Engineering High School; Ryokichi Sagane, Chihiro Ishii and Hukutaro Simamura graduated from the department of physics at Tokyo Imperial University; Toshio Amaki majored in electrical engineering at Waseda University in Tokyo; Masafumi Inoki studied physics at the Tokyo Engineering University; Fumio Yamasaki, Yataro Sekido, Yukio Miyazaki, and Isao Miura came from Hokkaido Imperial University; and Torao Ichimiya majored in physics at Osaka Imperial University. Both groups enjoyed Nishina's nonauthoritarian manner and the Laboratory's atmosphere of free discourse. Sekido, who entered the Nishina Laboratory in 1936, was one of the many who appreciated Nishina's receptivity to new ideas:

> One day in 1936, Nishina was reading *Comptes Rendus* when I opened the door of his room. He told me excitedly, "Auger is saying that there are soft and hard components in cosmic rays." In December 1936 I read at our colloquium G. Pfotzer's paper on his observation in the stratosphere. Nishina said instantaneously, "This is interesting. Let us try it." . . .
>
> In the spring of 1937, I read at our colloquium J. Barnothy and M. Forro's papers (1936, 1937). They reported that most of the charged particles at 732 mwe [*sic*] underground were so soft that half of them were absorbed by 1.5 cm Pb and thought the cosmic rays penetrating down to deep underground to be some neutral particles, probably neutrinos. When I finished my talk, Nishina said, "This is interesting. Let us try it at Shimizu tunnel."[23]

One major difference between the theory group and the cosmic ray group was Nishina's greater and more direct involvement in cosmic ray research. Nishina

FIGURE 5.5 Many photographs taken during the 1930s indicate that Nishina actively participated in cosmic ray research. Above: Nishina sitting in front of the large magnet for the cloud chamber. (Courtesy of the Special Collections, NCSU Libraries.) Below: Nishina controlling the counters. (Courtesy of AIP Emilio Segrè Visual Archives.)

coauthored more papers about cosmic rays than theoretical questions and, although his construction of two cyclotrons starting in 1936 allowed him little time, many photographs and reminiscences show that he continued to pay serious attention to cosmic ray research. Unlike the theory group, whose management and leadership Nishina delegated to Tomonaga, in the cosmic ray group, Nishina himself was the leader.

Nishina divided the cosmic ray group into two subgroups, each of which used different types of instruments. The first subgroup, comprised by Ishii, Sekido, and Miyazaki, used an ionization chamber to measure cosmic ray intensity. This group

FIGURE 5.6 Left: Nishina's cosmic ray group preparing the experiment at the Shimizu tunnel. Right: Nishina and Yoshio Fujioka inspecting tubes of Geiger–Müller counter. (Courtesy of the Institute of Physical and Chemical Research.)

worked on altitude dependence, "making measurements at mountain altitude, at sea level, and in the Shimizu tunnel, as deep as several hundred meters underground," to a point "lower than anyone else descended until 1960."[24] They used Steinke and Neher types of ionization chambers given to them by the inventors, and built and modified several of their own models in Riken's workshop. By the early 1940s, the Nishina type of ionization chamber, called *Nishina Ichigata*, was widely being used in various studies.

The other subgroup, comprised by Sagane, Takeuchi, and Ichimiya, built a cloud chamber with a counter control system and tried to take photographs of traces of particles, such as the mesotron. At Nishina's insistence, the group built a large cloud chamber with a diameter of 40 cm. Because Riken did not own a large DC electric source suitable for the purpose, Nishina gained permission to use a DC source used for charging submarine batteries at the Yokosuka Navy Yard. Then he sent Takeuchi and Ichimiya for a year "to take cosmic ray pictures":

> While they were analyzing these photographs, the papers of the Blackett group began to arrive in Japan. Nishina-san then asked the Navy to extend the time of use for this battery charger in order to take more cosmic ray pictures.[25]

Among the photographs they found one with a clear trace of mesotrons in 1938.

In the fall of 1935, Henry Victor Neher, a Caltech specialist in cosmic ray research, visited Japan to deliver his electroscope and also to give lectures at Riken (Figure 5.7). His journal provides a vivid picture of cosmic ray research and other activities in the Nishina Laboratory's early days:

> Tuesday morning [November 5, 1935] was spent preparing for the talk to be given at the Institute of Chemical and Physical Research by invitation from Dr. Nishina.
>
> I had taken one of the scopes [sic] along and had connected it up so that anyone could see it who liked. A rather long and complete outline was put on 4 large blackboards as most of these Japanese can read English. A rough reproduction of our iso-cosmic ray

FIGURE 5.7 Above left: Henry Victor Neher was a cosmic ray specialist at Caltech. Above right: Millikan and Neher working with their self-recording ionization chamber. Neher visited Japan in 1935 to deliver his electroscope to Nishina, and it had been widely used by the Nishina's cosmic ray research group. (Courtesy of the Archives, California Institute of Technology.) Below: the Neher type electroscope that was employed to detect the intensity of radioactivity in Nagasaki in August 1945. (Courtesy of the Institute of Physical and Chemical Research.)

> lines at sea level was put on the board. After the lecture, Nishina asked if they could copy the original. I could see no good objection.
>
> Steinke has sent one of his cosmic ray apparatus to Nishina. It is now working on the top floor of the building, but under a 6″ concrete ceiling. They plan to install it on the roof in a thin [*sic*] roofed shelter. Steinke's apparatus is quite complicated, and is fit only for a permanent station. The ionization chamber is 10 l. capacity filled with Co at 30 [*sic*] atm. It is cylindrical in shape and lies down. A pressure gauge tells if there are leaks. Nishina said they had been bothered with leaks. A Lindemann electrometer needle

> is photographed at regular intervals on a movable photographic plate. A compensating rheostat with 100 segments operated by a motor can be run at different speeds depending on the amount of radiation. These steps cause jumps of the needle on the electrometer which are undesirable. Steinke has since made a condenser compensator, but since he already has 6 or 8 stations going with the older apparatus, he does not want to change.
>
> The Japanese are great imitators. This fact is shown well in their copying of Compton's apparatus. Apparently it is a verbatim copy. They have, however, added a recording and automatic calibrating device. This apparatus they had on top of Mt. Fuji this summer. (By the way, the director of the Japanese Meteorological Service told me there is a stone house on top of Fuji where 6 men are stationed all year long. Reliefs go up once a month.) Their results agree with those of Compton at similar magnetic latitudes. Nishina said they had not determined the residual ionization.
>
> A committee on cosmic rays has been formed and the program consists in establishing 5 permanent stations. (1) Formosa at 5000–6000 ft at the meteorological station there, (2) on Mt. Fuji, (3) in Tokyo, (4) on one of the northern islands at sea level, and I think the (5) on one of the southern islands of the Japanese group, i.e., between Tokyo and Formosa. They plan on copying Steinke's apparatus for this purpose . . . .
>
> Nishina said they are planning on making a 40 cm cloud chamber with a large magnet, borrowing a generator from the Navy department. At the present time, they have an 8″ chamber in a 4000–5000 gauss field. He showed me several pictures taken of aluminum having been bombarded with alpha particles, giving off + electrons. They have about 0.4 gm of radium available for experimental use.[26]

Researchers in the Nishina laboratory frequently employed Neher's electroscope because it was portable and easy to manage. Later, in August of 1945, Japanese scientists used it to detect the intensity of radioactivity in Nagasaki after the Fat Man atomic bomb was dropped there.[27]

Equipped with better apparatus, the cosmic ray group produced notable results from 1936 on. The first and most important of these was the September 10, 1936 measurement of cosmic ray intensities in the Shimizu tunnel at a depth equivalent to 800 m of water.[28] Nishina and Ishii carried out several measurements at different positions and depths "to find the penetration of cosmic rays through various thickness of rocks, which consist mainly of diolite, the average density being probably 2.8." Cosmic rays penetrated much deeper than the physicists initially expected, definitely deeper than 800 m of water, at which the cosmic ray intensity was assumed to be zero. They employed Neher's electroscope, "kindly supplied" by Millikan, and communicated the results to Neher.

In early 1936, Nishina made a plan to determine whether the upcoming solar eclipse of June 19, 1936 would influence cosmic ray intensity.[29] The researchers chose two different sites to take their measurements. The first of these was Mt. Syari in Hokkaido, at which point the eclipse was total. The second was the roof of Riken, from which the eclipse was partial (78% of coverage). At Mt. Syari, the researchers carried about 800 kg of heavy apparatus and accessories to an observation hut located at an altitude of 1260 m above sea level. This equipment included Neher's electroscope and an ionization chamber 30 cm in diameter and 50 cm in length that had been constructed in Riken's workshop. The area "was everywhere covered with snow."[30] At the Tokyo observation post, the researchers took measurements using only Riken's ionization chamber. At both observation posts, the data indicated that "during the period of the eclipse," as compared to other days, no large variations occurred in intensity or fluctuation of either cosmic rays or of local radiations. "These fluctuations seem[ed] to have been a little smaller during the eclipse than the average."[31]

In 1937, Nishina, Takeuchi, and Ichimiya reported their efforts to experimentally test whether the soft components of cosmic rays at sea level were composed of "cosmic-ray electrons" as well as the protons.[32] For this purpose, they mounted lead bars that were 1.5 cm thick (and later 3.5 cm thick) in the middle of their 40-cm diameter cloud chamber and applied a magnetic field of about 17,000 oersteds. At sea level near Tokyo, they found, approximately 10 to 20% of cosmic-ray particles consisted of electrons and positrons. The rest were heavy particles:

> of both signs, which have much greater penetrating power for lead than protons of the same momentum ($H\rho$) would have. The specific ionization of some tracks is also much smaller than that of protons of the observed $H\rho$. *These results can most naturally be explained, if one assumes the existence of new particles of a mass heavier than that of an electron and lighter than that of a proton.* At about this time we received the paper of Street and Stevenson and then that of Anderson and Neddermeyer and saw that these authors had obtained similar results. Crussard and Leprince-Ringuet also recognized the existence of particles, which lose less energy through matter than expected for electrons on the theory of showers and produces smaller specific ionization than protons of the same $H\rho$ . . . .
>
> Until now we have obtained only one track which can probably be used for the determination of the mass. The initial value of $H\rho$ of the particle was 7.4×10^5 gauss-cm, after passing through lead it became 4.9×10^5 gauss-cm, showing the loss of about a half of the energy. The loss of energy by ionization and the range in lead calculated from the thickness of the lead bar and the final $H\rho$ are consistent, *if we assume the mass in question of the particle to be 1/7 to 1/10 that of the proton.* The above value $H\rho$ of and the specific ionization shown by the corresponding tracks are in accordance with the assumed mass . . . .
>
> Although the exact determination of the composition of the penetrating component of cosmic-ray particles has thus not yet been possible, its large part no doubt consist of the above new particles, through the existence of which various difficulties in connection with cosmic-ray phenomena, e.g., ionization, radiative effect, penetrating power, etc. now find a natural explanation.[33]

Nishina and his colleagues did not speculate about the meaning of their discovery, however. Although Nishina had known about Yukawa's theory of meson since 1934, his 1937 paper neither mentioned Yukawa's 1935 paper nor hinted of a link between his discovery and Yukawa's theory. It was Oppenheimer and Serber who first publicly recognized, in a note in the *Physical Review* published only one month after Neddermeyer and Anderson's paper, the possibility that the newly discovered particle in the cosmic ray might be what Yukawa had suggested.[34] A few months later, Anderson and Neddermeyer gave a name to the new particle: *mesotron*.[35] Were Nishina and his fellow Japanese physicists too cautious to endorse Yukawa's wild idea before it was approved in the West? Nishina's conservative view of the new particle is revealed in his letter to Bohr of August 28, 1937:

> Another thing which may interest you is the study of comic ray particles. As I told you on the occasion of discussions at the University of Osaka, we have been studying by means of a large Wilson cloud chamber the energy loss of cosmic ray particles in passing through lead. In the course of the experiments, we found that there exist particles which at higher energies radiate less than electrons and at lower energies ionize less than protons. *This could most naturally be explained in our opinion by assuming the existence of particles, the mass of which is larger than that of the electron and smaller than that of the proton, although this is not the unique solution.* The same conclusion had also been given by Neddermeyer and Anderson, by Street and Stevenson, and by Crussard and Leprince-Ringuet. We are at present trying to obtain more exact value of the mass of the particle. I am sending you a copy of our letter to the Editor of the *Physical Review*.[36]

However, in their 1939 paper, Nishina's cosmic ray team conveniently abandoned this conservative view and even hinted that they had considered the connection between the new particle and Yukawa's theory early on: "*Since we published the results of mass determination of the mesotron, the existence of which had theoretically been foreseen by Yukawa*, we have been continuing the same experiments with the Wilson cloud chamber."[37] They reported their arrangement of the instrument as follows:

> A lead bar 5 cm thick was mounted in the middle of the chamber 40 cm in diameter, which is filled with air and alcohol vapor, and placed in a magnetic field of about 12,600 oersteds. The operation of the chamber was controlled by two Geiger-Müller tube counters mounted immediately above the chamber. The distance between the counters was about 15 cm. Above the counters was placed a lead block 20 cm thick.[38]

The result was a clear photo of a trace of mesotron taken in September of 1938 (Figure 5.8). They also calculated the mass of the particle with the consideration or relativistic value: the new result was, $M_m = (180 \pm 20)m$, where m is the mass of the electron.[39]

In 1940, Nishina, Sekido, and Simamura, together with H. Arakawa of the Central Meteorological Observatory, published a series of papers on the relationship between cosmic ray intensity and different air masses. The first of these papers considered measurements of cosmic ray intensities under five different conditions of air masses in Tokyo during 1937; the measured intensities differed noticeably with respect to air conditions.[40] In 1937, they confirmed these findings by measuring cosmic ray intensities in different parts of Tokyo under air masses associated with 13 different cyclones. In "Cosmic Ray Intensities and Cyclones," they reported that "the passage of a cold air mass tends to increase the cosmic ray intensity, while that of a warm air mass tends to decrease it."[41] Cyclones are seasonal phenomena, and they next attempted to eliminate the seasonal effect by taking measurements under air masses caused by 25 cases of anticyclones. Their results were similar to their previous findings: "the arrival of a cold air mass tends to increase the intensity of the hard component of cosmic rays, while that of a warm air mass tends to decrease it, in conformity with the instability of the meson."[42]

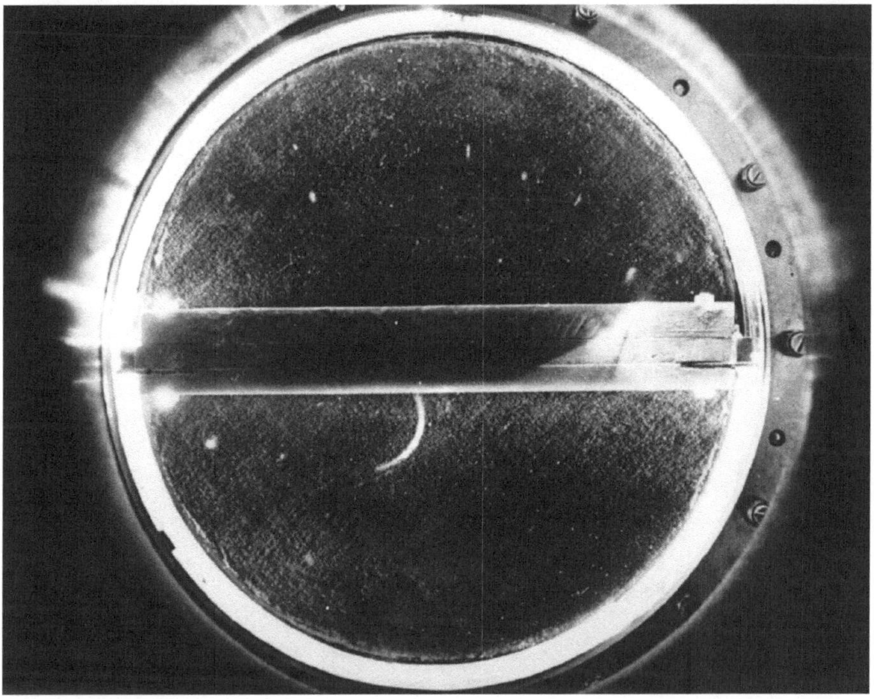

FIGURE 5.8 The track of a mesotron found by Nishina, Takeuchi, and Ichimiya, was published in their 1939 paper, "On the Mass of the Mesotron" *Physical Review*, *55* (1939), 585–586 on 585.

These results were contradicted, however, by measurements taken for a later study, in which they tested Donald H. Loughridge and Paul Gast's theory that cosmic ray intensities show noticeable changes in cold and warm fronts.[43] Although their results agreed well with Loughridge and Gast's outcomes for warm fronts, they failed to discover "any effect of the cold front beyond the statistical fluctuations." "All such differences," they suggested, probably had their "origins in the different structures of the upper atmospheres." Their further measurements of cosmic ray intensities during typhoons, which they reported in a 1941 paper, again confirmed the finding that cosmic ray intensities remained constant in cold air masses.[44]

In 1939, Nishina returned to the Shimizu tunnel to repeat the experiment that he and Ishii had done three years earlier, this time at a depth equivalent to 1400 m of water.[45] Under Nishina's guidance, Sekido, Miyazaki, Masuda, and two assistants spent "4600 hours of observation between September, 1939 and July, 1940" recording the number of coincidence hits caused by single rays and showers. Their results were inconclusive: they could not decide whether the primaries for the hard showers they observed were either "ionizing particles such as mesons or protons, or nonionizing particles — say neutrinos — as was proposed by Barnóthy and Forró."

The last study Nishina performed was in collaboration with Sekido, Miyazaki, and Karl Birus, a young German who had studied under Heisenberg at Leipzig.[46]

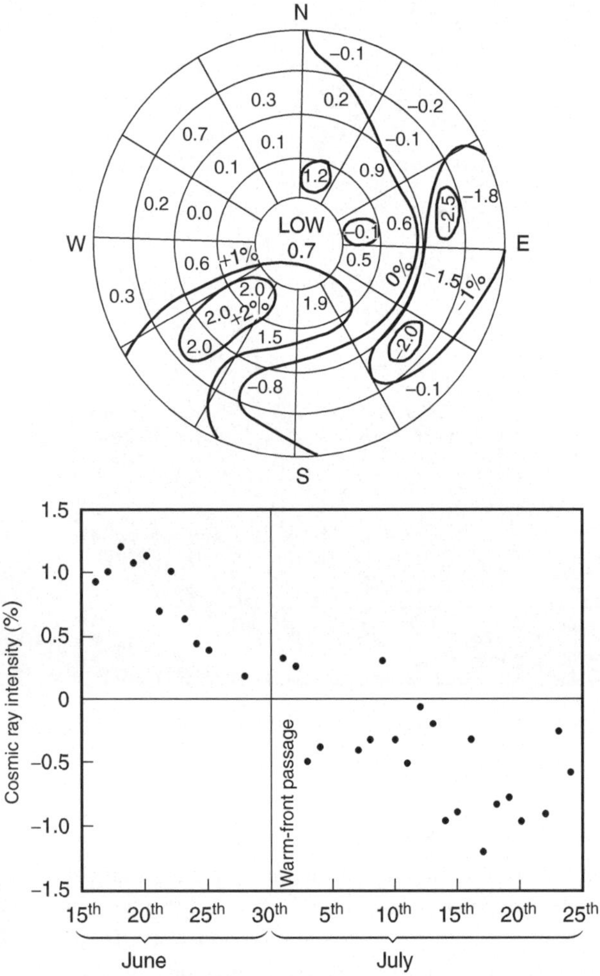

FIGURE 5.9 In the late 1930s, Nishina's cosmic ray group concentrated on the influence of different air masses on cosmic ray intensities. (Above: Y. Nishina, Y. Sekido, H. Simamura, and H. Arakawa, *Nature*, *145* [1940], 703–704. Below: Y. Nishina, Y. Sekido, H. Simamura, and H. Arakawa, *Physical Review*, *57* [1940], 1050–1051.)

To test Heitler's argument for the existence of a neutral meson, they aimed to detect it in cosmic rays. They used a cyclotron magnet as an analyzer for detecting the charge exchange of mesons ($P + Y^- \leftrightarrow N + Y^0$, $N + Y^+ \leftrightarrow P + Y^0$, where P is proton, Y meson, and N neutron). Their result was negative. Despite the importance of this finding, "the Boss hesitated to write an article,"[47] and then chose to send two papers on the subject to the *Scientific Papers of the Institute of Physical and Chemical Research* rather than more widely circulated journals like *Nature*, *Physical Review*, or *Zeitschrift für Physik*.

The cosmic ray group also carried out two long-term projects (Figure 5.10). In the first, they analyzed changes in cosmic ray intensity to determine how this intensity

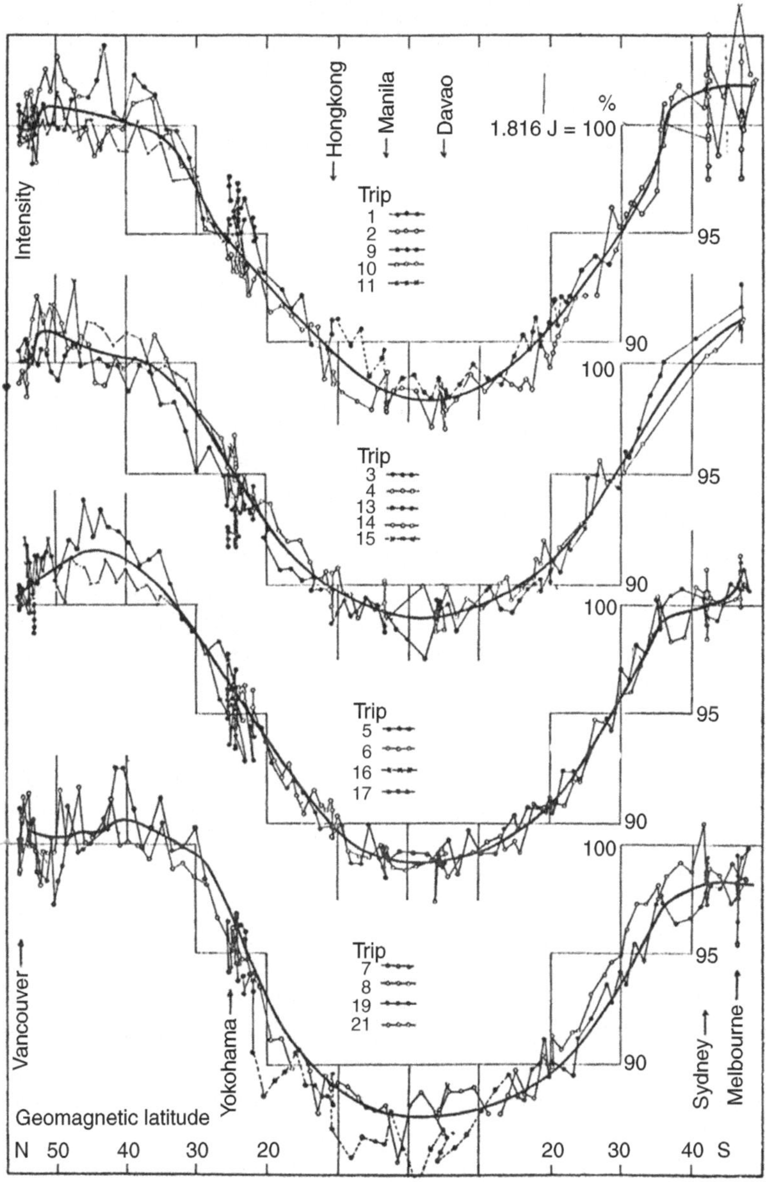

FIGURE 5.10 Nishina's cosmic ray group carried out two separate long-term projects to detect latitude and barometric effect on cosmic rays. Two Japanese ships that sailed from Japanese harbors to Melbourne and Seattle were employed for the purpose. (Y. Sekido, Y. Asano, and T. Masuda, "Cosmic Rays on the Pacific Ocean. Part I. Latitude Effect," *SP*, *40* [1943], 439–455 on 445.)

was affected by various factors: time, atmospheric pressure, temperature, moisture, air pollution, magnetic wind, sunspot activity, and even supernova.[48] Ishii, Sekido, Simamura, and Yoshiro Asano, used a standard Steinke apparatus to collect data between January of 1937 and July of 1938 and found that changes in cosmic ray intensity were closely correlated to changes in only two of these factors: time and atmospheric pressure.

The other long-term project was measuring the effects of latitude and barometric pressure on cosmic ray intensity.[49] Sekido, Asano, and Masuda set a Neher electroscope on board a ship, the *Kitano-maru* of the Nippon Yusen Kaisha (NYK) Australian line, which made eight voyages between Yokohama and Melbourne from April of 1937 to March of 1938. They set similar equipment on board another ship, the *Heian-maru* of the NYK Seattle line, during its 14 voyages between Kobe and Seattle from April of 1938 to April of 1939. After collecting the data, they carefully analyzed them and considered every possible source that might influence the observed latitudinal and barometric effects. They found that cosmic ray intensities were greatly influenced by the effects of magnetic latitude and geographic latitude. It was not until 1943, however, that they published these results.

In 1941, Nishina, Sekido, Takeuchi, and Ichimiya published a textbook, *Uchusen* (*Cosmic Rays*), which nicely summarized previous work in cosmic ray physics both in the West and in Japan (Figure 5.11).[50] In six chapters, they covered the historical background, the instruments used for cosmic ray measurement, the geophysical methods to study cosmic ray intensities, the laboratory research on cosmic ray intensities, the study of cosmic rays using tracks in the cloud chamber, and the future agenda of cosmic ray research. Nishina and his colleagues filled most of the book's pages with experimental data describing changes in cosmic ray intensity with respect to height, depth, latitude, and longitude, but they repeatedly emphasized its importance to theoretical research. The future development of theoretical physics, they claimed in the preface, "depends on the results from cosmic ray research," in particular the data describing tracks of cosmic rays "in the cloud chamber."[51] The authors also intensively discussed certain theoretical issues. They devoted three sections to discussions of Yukawa's meson theory,[52] and included in their last chapter, "The Future Agenda," long theoretical discussions.

In 1944, Sekido published a book under the same title, *Uchusen* (*Cosmic Rays*), which he designed for a general readership.[53] In this book, he included discussions of a considerable number of works by Japanese researchers and many photographs of instruments they had constructed and used. This book was published during the last days of the war, and Sekido's lip service to Japan's strong nationalism and patriotism is noteworthy. In Chapter 4, he attempted to illustrate a relationship between cosmic ray research and the war effort in five different areas: meteorology, geomagnetism, biology, the theory of matter, and cosmology. Although the cosmic ray group's research on meteorology and geomagnetism may have related to Japan's war effort, as Sekido claimed, their research touching on biology, the theory of matter, and cosmology did not. For example, in his discussion of cosmic ray research and the theory of matter Sekido included a long discussion of meson theory, which certainly had no connection with any war-related research.[54]

In May 31, 1944, Masahumi Ogawa and Ishii each submitted their papers to the *Bulletin of Institute of Physical and Chemical Research* that became the last

物 理 學

XI. B.

宇　　宙　　線

仁 科 芳 雄
關 戶 彌 太 郎
竹　內　　柾
一 宮 虎 雄

宇　　宙　　線

關 戶 彌 太 郎 著

科 學 新 書

52

FIGURE 5.11 Two textbooks on "uchusen (cosmic rays)" by Nishina's cosmic ray research group nicely summarized the recent development of cosmic ray research in the West and in Japan. Left: Y. Nishina, Y. Sekido, M. Takeuchi and T. Ichimiya, *Uchusen* (1941). Right: Y. Sekido, *Uchusen* (1944)

cosmic ray works before the end of the war. Ogawa used the cloud chamber to measure the mass of slow mesotrons in the cosmic rays, and suggested that there might be two kinds of mesotrons — each with 120 and 250 times heavier than the mass of electron.[55] Ishii's paper was on the details of the *Nishina Ichigata*, a cosmic ray meter that continuously measured the intensity of the hard components of cosmic rays to detect any change over time (Figure 5.12).[56] The measurements of cosmic ray intensities at high altitudes yielded by this instrument were so accurate that Ishii boasted that its margin of error was only "0.56%." With the aid of the Tokyo Electric Company, Riken's workshop constructed five of these instruments by the end of 1941, and all were in full operation by March of 1942. In 1944, as Japan's wartime operations deteriorated, the Nishina Laboratory moved all five of these meters to the Tokyo Observatory for use in collecting meteorological data on such phenomena as geomagnetic bursts.[57]

The cosmic ray research performed in the Nishina laboratory was part of the larger worldwide effort to determine the nature of these rays. Before determining their own agenda, the cosmic ray group carefully examined the experimental and theoretical work of Western researchers. The topics the group chose were the penetrating power of cosmic rays and the effects on cosmic rays of longitude, latitude, and air pressure. In these areas they achieved reasonable success. Their improvements of experimental instrumentation, such as their development of the *Nishina Ichigata* cosmic ray meter, also were important additions to cosmic ray research.

However, the cosmic ray research group never managed to be *first* in providing crucial experimental data or new theories to explain them. They repeated what Western

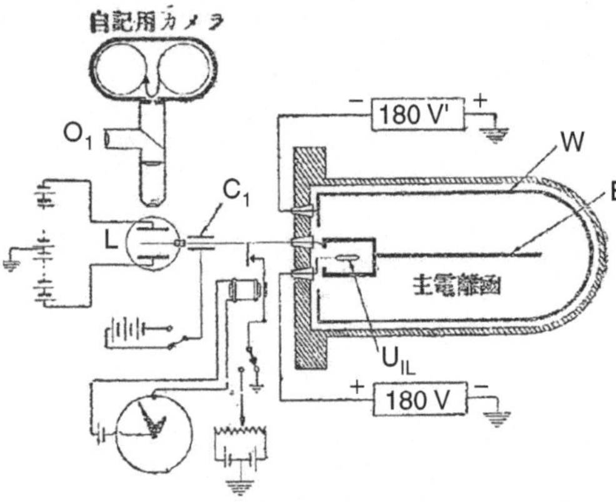

第1圖 宇宙線計概念圖

FIGURE 5.12 A late version of the *Nishina Ichigata* cosmic ray meter. This Japanese instrument very accurately measured cosmic ray intensities at high altitudes. (C. Ishii, *Bulletin IPCR*, *23* [1945], 191–201.)

researchers already had done, although the data they provided often were more accurate. Japanese researchers sometimes made discoveries almost simultaneously with their Western colleagues, but their papers, although solid enough for publication in elite Western journals like *Nature* or *Physical Review*, never were quite good enough to establish priority. They came close. In their 1937 paper, "On the Nature of Cosmic Ray Particles," Nishina, Takeuchi, and Ichimiya suggested the existence of an unknown particle heavier than the electron, but failed to connect it to Yukawa's 1935 theory of the meson that had predicted it.

Taken as a whole, the achievements of the Nishina Laboratory's cosmic ray group were less glamorous than those of the theory group. Nevertheless, their pre-1945 work was important. First, their research prepared common ground on which young Japanese experimentalists and theoreticians were able to work together. Without the cosmic ray group, the development of meson theory could not have been nearly as successful. Second, their work greatly contributed to the rapid postwar development of cosmic ray physics and astrophysics in Japan. Interest in cosmic rays and related subjects spurred the development of strong, competent research groups in the Japanese physics community, and members of the cosmic ray group trained the next generations of Japanese physicists, including Masatoshi Koshiba, cowinner of the 2002 Nobel Prize in physics "for [his] pioneering contribution to astrophysics, in particular for the detection of cosmic neutrino."[58]

5.4 COSMIC RAY RESEARCH AND THE MESON THEORY: COOPERATION BETWEEN EXPERIMENTALISTS AND THEORETICIANS

In the pre-1945 Japanese academic community, cooperation among scientists was rare. The Japanese university system included an institution called *koza* (the chair), which guaranteed a senior professor exclusive control of his field at his university. Assistant and associate professors, as well as the full professor's assistants, were required to faithfully and exclusively follow the professor's line of research. Even in a relatively small field like physics, this tradition was strictly observed. Cooperation between experimentalists and theoreticians, a common practice in the Cavendish Laboratory, Niels Bohr's Copenhagen Institute and American laboratories, was unthinkable.

The Nishina Laboratory in Riken provided an oasis from this tradition of academic isolationism. From the Laboratory's inception, cooperation and free dialogue between the cosmic ray and the theory groups was frequent and encouraged. Because Nishina himself worked in both areas, his subordinates felt free to practice in both areas. In 1980, two experimentalists, Takeuchi and Ishii, and two theoreticians, Tomonaga and Tamaki, had an opportunity to revisit the close interactions between the two research groups during the 1930s and 1940s when they were interviewed by Laurie M. Brown and four of his Japanese colleagues, Yoichi Fujimoto, Michiji Konuma, Ziro Maki, and Tetsuo Tsuji. Excerpts from that interview are revealing:

> **Brown**: I would like to ask something about the connection between theoretical work and experimental work during this time, especially at Riken. For example, when you needed some experimental information, did you consult the experimentalists?
>
> **Tomonaga:** Of course, I was very intimate with the experimentalists. In the seminar at Riken I was always with these two experimentalists — Ishii-san and Takeuchi-san. Sometimes I might have asked them to do very difficult or even impossible experiments.
>
> **Takeuchi:** No, Tomonaga-san did not ask us to do such difficult experiments but we were frequently asked to do experiments impossible at the time by the Boss (Dr. Nishina). For example, he wanted us to experiment on meson scattering, but to obtain accurate data we would have had to enlarge the cloud chamber itself by a great deal.

FIGURE 5.13 In the Nishina Laboratory, unlike elsewhere in the Japanese physics community, the cosmic ray and the theory groups often worked together. This photograph shows experimentalists of the cosmic ray group and theoreticians of the theory group in the Nishina Laboratory. Some researchers from the Takamine Laboratory also joined the excursion to Mt. Fuji to measure the intensity of cosmic rays. First row from the left: Y. Fujioka (Takamine Laboratory), F. Yamasaki, S. Tomonaga, M. Kobayasi, Y. Nishina, K. Tamaki, and M. Takeuchi. (Courtesy of the Institute of Physical and Chemical Research.)

. . .

Fujimoto: Were the experimental data accurate enough to compare with calculation results?

Tomonaga: Research on the intensity vs. depth curve was being done extensively by Ishii-san and his group, and they saw very clearly the change of slope. The curve, furthermore, was very consistent with foreign results. Thus the theoretical interpretation was an interesting subject both for theoretician and experimentalists.

. . .

Brown: This is a good example of theoretical–experimental collaboration . . . . I think this question is very important if we try to understand why Japanese physics was successful. It was more so in theory than in experiment, but I personally doubt that the important theoretical work could have been done without experimental support. Through personal contact between theoretical and experimental physicists, the theoreticians can learn which experimentalists to believe and which experiments are credible. One of my teachers, Hans Bethe, told me that this was a very important aspect of his career. Many people have said that in the case of France, where theorists and experimentalists were widely separated, physics declined.

. . .

Tomonaga: What interested me in the results from the Shimizu Tunnel Station were the results obtained by Ishii-san and others on cosmic ray bursts, i.e., possible multiple production [of mesons]. Unfortunately, only a limited number of data were available. I used to see these data several times per month. Do you remember, Ishii-san?
Looking at these data I wanted to know such-and-such information and asked the experimentalists, but regrettably, there was not enough data to extract statistically meaningful conclusions. The number of events deep underground was small, and furthermore we were interested in the even rarer multiple production events.

. . .

Fujimoto: My impression is that in those days you tried to do too many things at the same time.

Takeuchi: It was inevitable, because we had to compete with foreign groups that continued to publish results, many of which were also being obtained by Japanese groups.

Tomonaga: Fujimoto always claims that is so, but if we confined ourselves to small-scale experiments, I cannot believe they would have been better. A possible counter-example is the case of Fermi. He did not use accelerators but only water-filled basins and buckets to do his experiments. He made great achievements; his work was epoch-making. You may think that such experiments could have been done also in Japan. But I want to defend Japanese experimentalists: Fermi's group used a large amount of radium to produce their neutrons.

. . .

Tomonaga: But Fermi obtained very many results in a really short time. We would have had to do it immediately [to be competitive] but it was terribly difficult. Japanese experimentalists should not be blamed. The fault is the absence of theoreticians like Fermi in Japan.

Takeuchi: Tomonaga-san kindly defended Japanese experimentalists, but they were not quick enough. When I was asked to build a large cloud chamber, it might have been finished earlier if I had looked after each necessary part by myself. Instead, each piece was ordered from an appropriate company, and as a result there were delays in its completion. On this point foreign researchers had a superior power of action. Or else, it may be that the system is a more convenient one.

. . .

Maki: I heard once about a "fluffy meson." What is that?

Takeuchi: It is connected with a theory of Tamaki-san, High energy primary cosmic rays can produce many low energy mesons. This can appear as something like a number of fluffs around the tracks of a cosmic ray particle, so it was called "fluffy mesons."

Tamaki: Later Teketani used this idea in his theory of neutral mesons.

Konuma: As the fluffs were surely charged mesons, what did this have to do with neutral mesons?

Fujimoto: There were discussions about the altitude dependence of the soft component, and it turned out to be impossible to explain all of the soft components in the upper atmosphere as originating from decay electrons of mesons. Something new was needed, so they invented the fluffy mesons. At first it was not directly connected with neutral mesons.

Maki: Were there any early attempts to observe the neutral mesons?

Tomonaga: Yes, there was a work by Nishina and Karl Birus.

Tamaki: That was to detect the charge exchange of mesons [a meson giving up its charge on scattering from a nucleus], using the cyclotron magnet as an analyzer.[59]

Cooperation between experimentalists and theoreticians in the Nishina Laboratory could not be described as entirely fair to the experimentalists because the theoreticians by far had the best of the bargain. Although the theoreticians seldom utilized the experimental results by cosmic ray researchers in the same laboratory, nonetheless they had in-house specialists whom they could consult whenever it was necessary.[60] No theories suggested by Japanese physicists seem to have inspired cosmic ray experimentation in the Nishina Laboratory. Work by Western physicists, not Japanese theoreticians, inspired Nishina's frequent comment, "This is interesting. Let us try it." Surprisingly, Yukawa's theory of meson does not seem to have engendered Nishina's experimentation that resulted in a discovery explainable by a particle heavier than an electron but lighter than a proton. Instead, Nishina's discovery seems to have been purely experimental like that of Anderson and Neddermeyer.

Interestingly, Anderson's memoir strongly suggested that Yukawa's 1935 paper could have led experimentalists to discover the mesotron in cosmic rays earlier than 1937.

> We saw previously how the Dirac theory predicted the existence of positrons, although it played no role in their discovery. The discovery of mesons, similarly, was based on experimental measurements and procedures, with no guide from any theoretical predictions . . . .
>
> This novel suggestion of Yukawa's was unknown to the workers engaged in the experiments on the meson until after the meson's existence was established. Although Yukawa's suggestion preceded the experimental discovery of the meson, he published it in a Japanese journal which did not have a general circulation in this country. *It is interesting to speculate on just how much Yukawa's suggestion, had it been known, would have influenced the progress of the experimental work on the meson. My own opinion is that this influence would have been considerable even though Dirac's theory, which was much more specific than Yukawa's, did not have any effect on the positron's discovery.*

> *My reason for believing this is that for a period of almost two years there was strong and accumulating evidence for the meson's existence, and it was only the caution of the experimental workers that prevented an earlier announcement of its existence. I believe that a theoretical idea like Yukawa's would have appealed to the people carrying out the experiments*, and would have provided them with a belief that maybe after all there is some need for a particle as strange as a meson, especially if it could help explain something as interesting as the enigmatic nuclear forces.[61]

Indeed Nishina was the only scientist in the world who could have connected Yukawa's meson theory with the new heavy particle in cosmic rays before 1937. It was well within the realm of possibility for him to have gained priority in the discovery of the mesotron before Anderson and Neddermeyer achieved this feat in 1937. Nishina's team also seems to have had a photo of a track of mesotron before Street and Stevenson published their first photos of mesotron traces in the fall of 1937 in their article in *Physical Review*, "New Evidence for the Existence of a Particle of Mass Intermediate Between the Proton and Electron."[62] As Kobayasi remembered:

> One day in 1937, Nishina noticed that a singular track in a cloud chamber picture taken by M. Takeuchi might be a meson. He asked me if the mass could be obtained. I estimated it by using the formula of energy loss by ionization and concluded that the particle was certainly lighter than a proton and heavier than an electron. Nishina reported this at a meeting in Hokkaido.[63]

Nonetheless, Nishina neither mentioned Yukawa's theory in his 1937 paper nor hinted that his discovery was related to the theory. Nishina connected the new heavy particle with Yukawa's prediction only later, in a paper he published in 1939, after Western physicists had already seriously considered Yukawa's meson theory as the most plausible explanation for the mesotron.

Why was Nishina so hesitant to connect Yukawa's theory with his own discovery, despite his early endorsement of Yukawa's theory? It seems clear that, above all, Nishina, like many other Japanese theoreticians, did not fully appreciate the importance of Yukawa's theory. His continuous encouragement of Yukawa and his praise of Yukawa's work in 1934 seem to be the result of his habit of encouraging promising youngsters, not from a particular appreciation of the real impact of Yukawa's theory. Also, as a disciple of Bohr, Nishina was too conservative to wholeheartedly embrace a new, undiscovered particle. This conservative attitude no doubt influenced the other researchers in his laboratory between 1935 and 1937, when neither the theoreticians nor the experimentalists paid any serious attention to Yukawa's theory. For example, when Yoshio Fujioka, one of Tomonaga's patrons at Riken, asked Tomonaga what he thought of Yukawa's meson idea, Tomonaga merely responded, "Not too bad!"[64]

During this time, cooperation between the two groups had taken root, but it had not come close to the maturity and efficiency of cooperation between scientists in the West. All of the researchers in the Laboratory had been brought up in an academic tradition in which cooperation between two distinct research groups was nonexistent. In addition, except for Nishina, who had spent more than seven years in the top institutes like the Cavendish and Bohr's Institute, no member of either group had any experience working together with researchers from other groups. Cosmic ray research was the common project that provided these young scientists with their first opportunity to learn a new, more efficient way of conducting research. The importance of the Nishina Laboratory's new cooperative effort that embraced both experimentalists and theoreticians is that it began a new tradition in the Japanese science community, a tradition that led to spectacular success in the field of elementary particle physics in Japan. It was Nishina who brought this cooperation into existence.

NOTES

1 H. Victor Neher, "Some of the Problems and Difficulties Encountered in the Early Years of Cosmic Ray Research," in Yataro Sekido and Harry Elliot (eds.), *Early History of Cosmic Ray Studies: Personal Reminiscences with Old Photographs* (Dordrecht: D. Reidel Publishing Company, 1985), pp. 91–97 on p. 91.

2 "Nuclear Research at RIKEN: Dialogue with the Late Sin-itiro Tomonaga," in Laurie M. Brown, et al. (eds.), *Particle Physics in Japan, 1930–1950*, Vol. II (Kyoto: Research Institute for Fundamental Physics, Kyoto University, 1980), pp. 1–25 on p. 14.

3 Y. Nambu, "Summary of Personal Recollections of the Tokyo Group," in Laurie M. Brown, et al. (eds.), *Elementary Particle Theory in Japan, 1935–1950* (Kyoto: Research Institute for Fundamental Physics, University of Kyoto, 1988), pp. 3–6 on p. 6.

4 Victor F. Hess, "The Discovery of Cosmic Radiation," *Thought (Fordham University Quarterly), 15* (1940), 225–236.

5 Victor F. Hess, "Über Beobachtungen der durchdringenden Strahlung bei sieben Freiballonfahrten," *Physikalische Zeitschrift, 13* (1912), 1084–1091.

6 Victor F. Hess, "Unsolved Problems in Physics; Tasks for the Immediate Future in Cosmic Ray Studies," in B. Samuelsson and M. Sohlman (eds.) *Nobel Lectures, Physics 1922–1941* (Amsterdam: Elsevier Publishing Co., 1965), pp. 360–362. Italics are added.

7 Robert A. Millikan and I. S. Bowen, "High Frequency Rays of Cosmic Origin, I. Sounding Balloon Observations at Extreme Altitude," *Physical Review, 27* (1926), 353–361; R. A. Millikan and R. M. Otis, "High Frequency Rays of Cosmic Origin. II. Mountain Peak and Airplane Observations," *Physical Review, 27* (1926), 645–658; R. A. Millikan and G. Harvey Cameron, "High Frequency Rays of Cosmic Origin. III. Measurements in Snow-Fed Lakes at High Altitude," *Physical Review, 28* (1926), 851–868.

8 Robert A. Millikan, "The Last Fifteen Years of Physics," *Proceedings of the American Philosophical Society, 65* (1926), 78; R. A. Millikan, *Science and the New Civilization* (New York: C. Scribner's Sons, 1930), p. 105. See also Robert Kargon, "Birth Cries of the Elements: Theory and Experiment along Millikan's Route to Cosmic Rays," in Harry Woolf (ed.), *The Analytic Spirit* (Ithaca: Cornell University Press, 1981), pp. 309–325.

9 J. Clay, "Penetrating Radiation," *Proceedings of the Royal Academy of Amsterdam, 30* (1927), 1115–1127; "Der kosmische Ursprung der Höhenstrahlung," *Naturwissenschaften, 15* (1927), 356–357; "Penetrating Radiation. II," *Proceedings of the Royal Academy of Amsterdam, 31* (1928), 1091–1097; "Ultra Radiation (penetrating radiation). III. Annual Variation and Variation with the Geographical Latitude," *Proceedings of the Royal Academy of Amsterdam, 33* (1930), 711–718.

10 Dimitry V. Skobeltzyn, "The Early Stage of Cosmic Ray Particle Research," in Laurie M. Brown and L. Hoddeson (eds.), *The Birth of Particle Physics*, pp. 111–119. It was reprinted in Sekido and Elliot (eds.), *Early History of Cosmic Rays Studies*, pp. 47–52.

11 W. Bothe and W. Kolhörster, "Das Wesen der Höhenstrahlung," *Zeitschrift für Physik, 56* (1929), 751–777 on 777. See also Georg Pfotzer, "Early Evolution of Coincidence Counting: A Fundamental Method in Cosmic Ray Physics," in Sekido and Elliot (eds.), *Early History of Cosmic Ray Studies*, pp. 39–44.

12 Bruno Rossi, "Arcetti, 1928–1932," in Sekido and Elliot (eds.), *Early History of Cosmic Ray Studies*, pp. 53–73 on p. 72. He remembered: "So incredible were my results that a German magazine (if I remember correctly, it was *Naturwissenschaften*) refused to publish my paper. The paper was then accepted by *Physikalische Zeitschrift* after Heisenberg had vouched for my credibility."

13 Satio Hayakawa, *Cosmic Ray Physics: Nuclear and Astrophysical Aspects* (New York: John Wiley & Sons, 1969), p. 4.

14 Laurie M. Brown, "Nuclear Forces, Mesons, and Isospin Symmetry," in Larie M. Brown, et al. (eds.), *Twentieth Century Physics*, Vol. 1, pp. 357–419 on p. 385.

15 M. Takeuchi, "Cosmic Ray Study in Nishina Laboratory," in Sekido and Elliot (eds.), *Early History of Cosmic Ray Studies*, pp. 137–143 on p. 137.

16 Ibid., 137–138.

17 Y. Nishina, R. Sagane, and M. Takeuchi, "Research on Cosmic Rays Using the Wilson Cloud Chamber (preliminary)," *Bullettin IPCR, 12* (1933), 1014–1015.

18 For the establishment of the society, see *The Annual Report of the Japan Society for the Promotion of Science, 1* (1932), pp. 2–3; and Oh, *Nishina Yoshio and the Modern Physics in Japan*, pp. 179–183. The social background is offered in T. Hirosige, "Social Conditions for Prewar Japanese Research in Nuclear Physics," in S. Nakayama, et al. (eds.), *Science and Society in Modern Japan*, pp. 202–220 on pp. 206–210.

19 Oh, *Nishina Yoshio and the Modern Physics in Japan*, p. 182.

20 *Japan Society for the Promotion of Scientific Research, Outline of Research Projects under Special Committees and Subcommittees, No. 4* (1939), 167.

21 Yuzuru Watase, "Kikuchi Seishi Sensei and His Laboratory: The Foundation of Osaka Imperial University and Cosmic Ray Research [in Japanese]," *Shizen* (September, 1966), 28–31.

22 Hirosige, "Social Conditions for Prewar Japanese Research in Nuclear Physics," in S. Nakayama, et al. (eds.), *Science and Society in Modern Japan*, pp. 214–217.

23 Yataro Sekido, "Intensity and Anisotropy of Cosmic Rays," in Sekido and Elliot (eds.), *Early History of Cosmic Ray Studies*, pp. 187–206 on pp. 192–193.

24 Roundtable Discussion, "Cosmic Ray Research in Japan before World War II," in Laurie M. Brown, et al. (eds.), *Particle Physics in Japan, 1930–1959*, Vol. 1, (Kyoto: Research Institute for Fundamental Physics, Kyoto University, 1980) pp. 23–42 on pp. 30–31.

25 Roundtable Discussion, "Nuclear Research at RIKEN," in Laurie M. Brown, et al. (eds.), *Particle Physics in Japan, 1930–1950*, Vol. II, pp. 1–25 on p. 14. For the extended use of the DC source, Masatoshi Okochi, the director of Riken, wrote the Admiral Furuichi Tatsuo, the director of the Yokosuka Naval Yard (ibid., p. 25).

26 *Journal of H. Victor Neher for his trip in 1935* (personal copy of Ms. Topsy Neher Smalley). Neher visited Japan on the way back home from Manila via Honolulu. He arrived in Yokohama on November 3 and stayed in Japan for four nights.

27 "Cosmic Ray Research with Ion Chambers and Counters," in Tamaki and Ezawa (eds.), *Nishina Yoshio*, (Tokyo: Misuzu Shobo, 1991) pp. 112–113.

28 Y. Nishina and C. Ishii, "A Cosmic Ray Burst at a Depth Equivalent to 800 m of Water," *Nature, 138* (1936), 721–722.

29 Y. Nishina, C. Ishii, Y. Asano, and Y. Sekido, "Measurements of Cosmic Rays during the Solar Eclipse of June 19, 1936," *Japanese Journal of Astronomy and Geophysics, 14* (1936–1937), 265–275.

30 Ibid., 275.
31 Ibid.
32 Y. Nishina, M. Takeuchi, and T. Ichimiya, "On the Nature of Cosmic Ray Particles," *Physical Review*, *52* (1937), 1198–1199.
33 Ibid. Italics are added.
34 J. R. Oppenheimer and R. Serber, "Note on the Nature of Cosmic Ray Particles," *Physical Review*, *51* (1937), 1113.
35 Carl D. Anderson and Seth H. Neddermeyer, "Mesotron (Intermediate Particle) as a Name for the New Particle of Intermediate Mass," *Nature*, *142* (1938), 878. It was Millikan who advised them to adopt the name for the new particle.
36 Y. Nishina to N. Bohr (August 28, 1937) in *Supplement to the Publications*, pp. 19–22 on p. 22. Italics are added.
37 Y. Nishina, M. Takeuchi, and T. Ichimiya, "On the Mass of the Mesotron," *Physical Review*, *55* (1939), 585–586 on 585. Italics are added.
38 Ibid., 585.
39 Ibid., 586.
40 Y. Nishina, Y. Sekido, H. Simamura, and H. Arakawa, "Cosmic Ray Intensities and Air Masses," *Physical Review*, *57* (1940), 663.
41 Y. Nishina, Y. Sekido, H. Simamura, and H. Arakawa, "Cosmic Ray Intensities and Cyclones," *Nature*, *145* (1940), 703–704.
42 Y. Nishina, Y. Sekido, H. Simamura, and H. Arakawa, "Cosmic Ray Intensities in Relation to Cyclones and Anticyclones," *Nature*, *146* (1940), 95.
43 Y. Nishina, Y. Sekido, H. Simamura, and H. Arakawa, "Air Mass Effect on Cosmic Ray Intensity," *Physical Review*, *57* (1940), 1050–1051. See also Donald H. Loughridge and Paul Gast, "Air Mass Effect on Cosmic Ray Intensity," *Physical Review*, *56* (1939), 1169–1170.
44 Y. Nishina, Y. Sekido, H. Simamura, and H. Arakawa, "Cosmic Ray Intensities and Typhoons," *Physical Review*, *59* (1941), 679.
45 Y. Nishina, Y. Sekido, Y. Miyazaki, and T. Masuda, "Cosmic Rays at a Depth Equivalent to 1400 Meters of Water," *Physical Review*, *59* (1941), 401.
46 Y. Nishina, Karl Birus, Y. Sekido, and Yukio Miyazaki, "Ein Umwandlungseffekt neutraler Mesotronen," *SP*, *38* (1941), 353–359; Y. Nishina and Karl Birus, "Neutrale Mesotronen in der Höhenstrahlung?" *SP*, *38* (1941), 360–370.
47 "Nuclear Research at RIKEN," in Laurie M. Brown, et al. (eds.), *Particle Physics in Japan, 1930–1950*, Vol. II, p. 21.
48 C. Ishii, Y. Asano, Y. Sekido, and H. Simamura, "Change of Cosmic Ray Intensity according to Time," *Bullettin IPCR*, *18* (1939), 1066–1087.
49 Y. Sekido, Y. Asano, and T. Masuda, "Cosmic Rays on the Pacific Ocean. Part I — Latitude Effect," *SP*, *40* (1943), 439–455; Y. Sekido, "Cosmic Rays on the Pacific Ocean. Part II — On the Barometer Effect of Cosmic Rays," *SP*, *40* (1943), 456–466.
50 Y. Nishina, Y. Sekido, M. Takeuchi, and T. Ichimiya, *Uchusen* [*Cosmic Rays*, in Japanese] (Tokyo: Iwanami Shoten, 1941).
51 Ibid., p. 1.
52 Ibid., Section 27 of Chapter 5 and Sections 34 and 35 of Chapter 6.
53 Y. Sekido, *Uchusen* [*Cosmic Rays*, in Japanese] (Tokyo: Kasui Shobo, 1944).
54 Ibid., pp. 213–233.
55 M. Ogawa, "Cloud Chamber Study of Slow Meson," *Bullettin IPCR*, *23* (1945), 202–206.
56 The paper was however published after the end of the War. C. Ishii, "Characters of 'Nishina Ichigata' Cosmic Ray Meter," *Bullettin IPCR*, *23* (1945), 191–201.
57 Ibid., 201.

58 Royal Swedish Academy of Science, "The Nobel Prize in Physics 2002," (October 8, 2002), http://nobelprize.org./nobel_prizes/physics/laureates/2002/.

59 "Nuclear Research at RIKEN," in Laurie M. Brown, et al. (eds.), *Particle Physics in Japan, 1930–1950*, Vol. II, pp. 1–25 on pp. 8–12, 16, and 22.

60 The only reference that the theoreticians employed for their papers is Nishina's 1937 paper on the discovery of heavier particle. See M. Kobayasi and S. Ozaki, "On the Energy Loss of Fast Charged Particles by Pair Creation," *SP*, *34* (1938), 321–331 on 331.

61 Carl D. Anderson, "Unraveling the Particle Content of the Cosmic Rays," in Sekido and Elliot (eds.), *Early History of Cosmic Ray Studies*, pp. 117–132 on pp. 126–127. Italics are added.

62 J. C. Street and E. C. Stevenson, "New Evidence for the Existence of a Particle of Mass Intermediate Between the Proton and Electron," *Physical Review*, *52* (1937), 1003–1004.

63 Roundtable Discussion, "Particle Physics in Japan in the 1940s," in Laurie M. Brown, et al. (eds.), *Particle Physics in Japan, 1930–1950*, Vol. I, pp. 43–67 on p. 49.

64 Ibid.

6 Father of Big Science: The Construction of Two Cyclotrons

> [Nishina's 26-inch cyclotron] is the number two in the world after the cyclotron in Dr. Lawrence' laboratory at the University of California Berkeley.
>
> **Asahi Shimbun (April 1937)**[1]

> This is ten years of my life . . . . It has nothing to do with bombs.
>
> **Y. Nishina (November 1945)**[2]

Among Nishina's many contributions to the Japanese physics community during the 1930s and 1940s, the most celebrated — and also the most controversial — were the construction and operation of his two cyclotrons at Riken. He was frequently seen working with one of these cyclotrons, as he is shown in a photo of him with his 60-inch cyclotron that was published on the cover of *Nishina Yoshio*, a collection of memoirs by his colleagues, students, and sons (Figure 6.1).[3] Nishina's cyclotron projects introduced Japan to the practice of *Big Science*, with its reliance on big staffs operating big machines in a big laboratory on big budgets.

What were Nishina's original objectives for his two cyclotrons, and what use did he actually make of them? How did he secure the enormous financial support needed to build and operate them? What difficulties did he encounter, and how did he overcome them? What contributions to Nishina's cyclotron projects were made by Ernest O. Lawrence and his associates at the University of California Berkeley? Why did Nishina move directly from his first cyclotron project of building a small 26-inch cyclotron to the considerably more ambitious project of building a large 60-in one, entirely skipping of constructing a medium-sized cyclotron (30–40-inch, 8–12 MeV)? Finally, how much did Nishina's cyclotron projects influence the future path of Japanese physics? These questions are investigated in this chapter.

6.1 THE SMALL (26-inch) CYCLOTRON

6.1.1 PRELIMINARIES

In the 1930s and 1940s, Japanese physicists constructed particle accelerators at Riken and the imperial universities in Tokyo, Osaka, and Kyoto. That Japanese physicists were able to gain financial support for these projects was an indication of the changing milieu in the 1930s. As Japan sank into the worldwide Great Depression from 1929, and particularly after Japan successfully invaded Manchuria

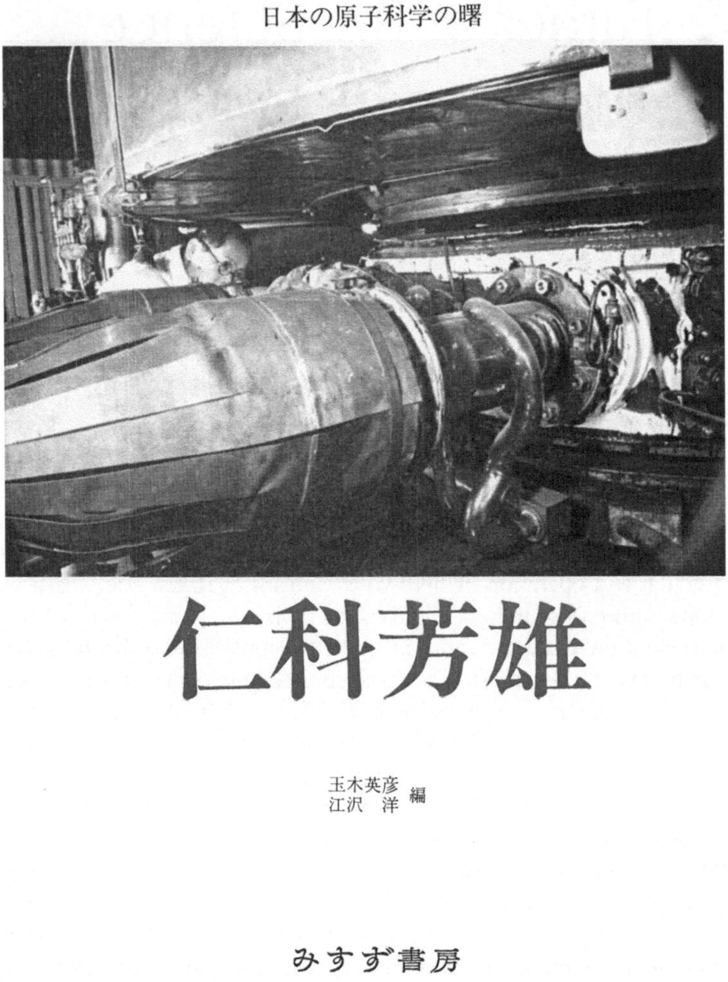

FIGURE 6.1 The cover *Nishina Yoshio* (Tokyo: Misuzu Shobo, 1991). Nishina was working with the 60-inch cyclotron.

in 1931, Japanese industrial and military leaders enthusiastically encouraged their nation's scientists and engineers to pursue an independently Japanese way of doing science and engineering.[4] Independence from Western science and engineering, they believed, would benefit Japan's industrial and military strength and demonstrate Japanese superiority to the rest of the world. Thus, the rise of Japanese militarism and chauvinism throughout the 1930s was largely responsible for a rapid development of science and engineering in Japan. As Japan's militarist expansion supported industry, Tetu Hirosige pointed out, industrial expansion supported science:

> The beginning of experimental nuclear physics depended largely upon funds donated by business circles. It is noteworthy that in the mid-1930s Japanese business circles

> were able to make such large contributions to a field like nuclear physics, which was not then expected to yield immediately any practical results. That is, big business' funding capability rested on a new stage in Japan's economy. Seizing the Manchurian Incident [the confrontation that led to the Japanese occupation of Manchuria] as an opportunity to escape from stagnation, Japan's economy entered a boom supported by colonial exploitation and a rapidly expanding war industry. It was this boom that allowed economic circles such latitude in making substantial grants to nuclear physics.[5]

When Nishina's first cyclotron was completed in April of 1937, an influential Japanese newspaper, *Asahi Shimbun*, proudly announced that the 26-inch cyclotron was "the number two in the world after the cyclotron in Dr. Lawrence's laboratory at the University of California Berkeley." The machine's purpose, the article asserted, was to solve "the riddle of the universe."[6]

Nishina's major sponsors for the construction of the two cyclotrons were the Japan Society for the Promotion of Scientific Research and the Mitsui Ho-onkwai Foundation. The Japan Society eventually provided more funding for nuclear research than for any other kind, even research more closely related to Japan's militarist and public health ambitions.[7] In 1937, for example, according to a careful study by Shizue Hinokawa, the budget of the Japan Society's 10th subcommittee for Research on Cosmic Rays and Nuclear Physics (which went mainly to fund construction of Nishina's 60-inch cyclotron) accounted for 24.1% of the combined expenditures of all of the projects supported by the Japan Society's more than 100 sub- and special committees plus those of the Geophysical Exploration Laboratory.[8] From 1934 to 1942, the 10th subcommittee's total expenditures for the cyclotron project amounted to 443,528 yen. The cyclotrons' other major sponsor, the Mitsui Ho-onkwai Foundation ("Ho-onkwai" means "repayment of kindness (to the nation)"), was established in 1934 by one of the largest conglomerates in Japan, the Mitsui zaibatsu. Many researchers acknowledged this foundation's support in their publications in the fields of cosmic ray physics, nuclear physics, and radioactive biology. Hinokawa explained the intense interest of the Mitsui zaibatsu in supporting nuclear research:

> Clearly, the zaibatsu capitalists of the day felt that cultural advancement was essential to enhancing and expanding the strength of the nation and that an organ was needed to encourage projects that benefited society and promoted culture. At the same time, enhancing and extending national strength necessarily meant developing and maintaining high engineering standards, which implied improving and expanding scientific research that could provide the foundations for technological advancement. The Nuclear Research Laboratory at IPCR [Riken] materialized under these circumstances.[9]

Thanks to Nagaoka, who was one of the foundation's 20 trustees as well as an influential member of the grant committee, the foundation awarded a three-year, 15,000 yen grant to Riken for "the artificial conversion of elements and study of their radiation" that, according to the Mitsui Ho-onkwai Foundation, had been delayed by its great expense but "must start immediately because of its great importance."[10] This funding speeded construction of the 26-inch cyclotron. The foundation later supported the construction of the 60-inch cyclotron through two grants of 10,000 yen each to the 10th subcommittee in 1939 and 1940.

Nishina took advantage of Japan's new interest in science and engineering to attract wider support and funding from both within and outside of Riken. Nishina believed, and was able to convincingly argue to potential funders, that with a cyclotron at their disposal the researchers in his laboratory would be in an excellent position to beat Westerners in the areas of nuclear physics and radioactive biology. The two fields were relatively new and having a cyclotron would put his laboratory on somewhat of an equal footing with the West. This belief explains, at least partially, how Nishina was able to garner such strong financial support for the construction of his two cyclotrons from the Japan Society, the Mitsui zaibatsu and other Japanese companies. Nishina also worked to gain public support for his efforts. Beginning in the mid-1930s, he published articles in popular science journals that introduced Japanese to recent developments in physics. He focused two of these articles on cyclotrons: (1) a 1939 review of various kinds of particle accelerators in which he gave a fairly detailed explanation of the cyclotron; and (2) a 1940 article about Lawrence in which he explained the principles of the cyclotron and its wide applicability to biology, medicine, and other disciplines.[11]

Officially, construction and operation of the cyclotrons were to be the tasks of a new laboratory at Riken, the Nuclear Research Laboratory, but it was clear from the beginning that Nishina was the true leader of the cyclotron effort and of the nuclear research facility. A close collaboration between the laboratories of Nishina and Shoji Nishikawa led to the opening of the Nuclear Research Laboratory in January of 1935. Its mission was "extensive investigation on nuclear physics, biology, and related problems."[12] The Mitsui Ho-onkwai Foundation, the Tokyo Electric Light Company, and the Japan Wireless Telegraph Company funded the Nuclear Research Laboratory, and additional grants-in-aid came from the 10th subcommittee of the Japan Society.[13] The physical premises of the new laboratory were very large, consisting of "six buildings, the total area covered being about 1225 m^2."[14]

The Nuclear Research Laboratory was organized in novel way: the research performed there was not the province of any single scientist, and its researchers crossed the lines of academic factions that had long dominated Japanese academic society. Riken's scientists from the laboratories of Nishikawa, Nishina, Takamine, Ishida, and Nagaoka worked in the Nuclear Research Laboratory alongside scientists from Tokyo Imperial University.[15] As Dong-Hoon Oh, Korean historian of science, observed:

> The opening of the Nuclear Research Laboratory was very important for Nishina. He could secure the support from senior scientists like Nishikawa, Ishida, Takamine and Nagaoka, and that support enabled him to attract more aid from the outside of Riken. Also, through this new laboratory, he could attract many talented researchers from other laboratories of Riken. For example, Nishikawa sent his own researchers like Tameichi Yasaki to Nishina in order to participate in the construction of the cyclotron.[16]

The two cyclotrons were not the only particle accelerators planned for the Nuclear Research Laboratory. Nishina and Nishikawa also considered simultaneously building a Cockcroft–Walton type accelerator. However, as Nishina wrote Bohr on August 3, 1936, some voices were raised against this plan:

> At present I am much occupied with the construction of a cyclotron and a high tension installation as well as [a] building for housing them. We hope to start the cyclotron at the end of October. The magnet, however, is 66 cm in diameter and 14,000 gauss in field strength and is not large enough. We are therefore now trying to get a fund for constructing a larger magnet. It is very interesting to see from your letter that you are also constructing a cyclotron as well as a high tension installation.
>
> I should like in this connexion [*sic*] to ask you one question. In our laboratory we have been discussing very much whether we should build a high tension source or not. The high tension which we can obtain by means of the Cockcroft method is 1 to 2 million volts or 3 millions at the highest as will be done at Cambridge. The energy obtained by such an installation is thus much less than that obtained by a cyclotron of say 66 cm in diameter and 17,000 [*sic*] gauss in field strength. Some of us therefore maintain that we should abandon the construction of the high tension source and use the money instead for the magnet of the cyclotron to be built and increase its diameter, that will be more useful. Some of us, however, contend on different grounds that we should construct the high tension source of the Cockcroft type. The settlement of this matter is very urgent in our case.
>
> Now I see from your letter that you are going to build a cyclotron as well as a high tension source. I should like very much to know for what purpose you are going to use the high tension source besides cyclotron.[17]

A decision was reached to build the controversial "high tension source," and the division of labor soon became evident: the Nishina Laboratory would be responsible for the construction of the cyclotron and the Nishikawa Laboratory would be responsible for the construction of the Cockcroft–Walton type accelerator.

Construction of the Cockcroft–Walton type accelerator started in 1936 with a group of researchers from the laboratories of Nishikawa, Nishina, and Nagaoka. This machine initially produced only 700 kV but eventually reached a production of 1 million volts, as originally planned. Later, the same group assisted physicists at Tokyo Imperial University in their construction of a 2-MV Van de Graaff accelerator. Researchers employed the Cockcroft–Walton type accelerator for investigations of artificial radioactivity, nuclear disintegration, γ-rays, and neutron scattering. Sugiura and Osamu Minakawa investigated neutron groups with 600 kV. Kenichi Shinohara, Mitio Hatoyama, Ziro Yuhara, and Takayuki Maeyama published a series of papers on γ-rays from fluorine that had been bombarded with protons with 1000 kV.[18]

6.1.2 The Machine

Construction of the small cyclotron started in the spring of 1936.[19] Nishina adopted Ernest Lawrence's 27-inchcyclotron at Berkeley as the prototype, partly because he had acquired an electromagnet of similar size (26 inch) and partly because he was fairly familiar with Lawrence's machine, which he had been studying since the early 1930s when he first communicated with Lawrence. Nishina and other Japanese researchers had acquired considerable knowledge about the Berkeley cyclotron from several papers that Lawrence and his collaborators had published about its design and operation in the *Physical Review* and other journals.[20] In April of 1932 he wrote to Lawrence:

> I am much interested in your work [on the production of high speed protons by the cyclotron] and should be very much obliged to you if you would send me a reprint, when you publish the detailed account of your experiments, for which I am working with a keen interest.[21]

Before starting construction of his first cyclotron, Nishina prudently sent two emissaries, Sagane and Yasaki, to Berkeley. Sagane arrived at Berkeley in the summer of 1935 and stayed for a year as principal liaison between the laboratories of Nishina and Lawrence. He faithfully accomplished his mission, delivering every bit of information he collected about the cyclotron at Berkeley to Nishina:

> [Sagane] did his job if anything too well: following his instructions, his compatriots copied Berkeley['s] mistakes as well as successes and Lawrence had to instruct them to fill up holes they had drilled in the tank plates in emulation of a measure once tried and immediately discarded at the Laboratory to improve the magnetic field.[22]

Yasaki stayed in Berkeley between September of 1935 and January of 1936 to gain some experience of the cyclotron. After returning to Japan, he continued to exchange letters with Lawrence and with Lawrence's instrument maker and confidant, Donald Cooksey, keeping them informed about the cyclotron construction process and sometimes asking for help solving technical difficulties. On one occasion, he asked Cooksey to send spare platinum foils that he could not get in Japan.[23] Nishinarepeatedly thanked Lawrence for accepting his two emissaries. On July 30, 1936, he wrote, "Our cyclotron is now under construction with the help of experience obtained by Yasaki in your laboratory as well as information from Sagane." On February 21, 1938, he wrote: "It is entirely due to the kind assistances and useful advices which you gave us through Sagane and Yasaki that we succeeded in the construction and the operation of our first cyclotron."[24]

Nishina's 26-inch cyclotron consisted of three parts: electromagnet, vacuum chamber, and high frequency system (Figure 6.2).[25] The huge electromagnet, donated by the JapanWireless Telegraph Company, originally had been used for a Poulsen arc generator at their Haranomati Station. Shibaura Engineering Works made necessary modifications in the poles and the coils of the magnet. The poles had a cylindrical shape with a diameter of 66 cm and gave rise to a magnetic field of 12,800 oersteds when driven by a current of 77 A at about 400 V. The vacuum chamber fit between the pole pieces with a gap of two 5 mm for shimming top and bottom. The 76 coils swam in two tanks each 120 cm in diameter. An automatic voltage regulator invented by Masakazu Takahashi was used to maintain the exciting current of the magnet constant within 0.1%. The total weight of the magnet assembly including the exciting coils and tanks was about 23 tons.

The design of the vacuum chamber was also "very similar to those of the 27½″ cyclotron at Berkeley."[26] Its outer diameter was 67 cm and its height was16.5 cm. To prevent leakage, sealing wax was applied to every possible part. Two metal oil

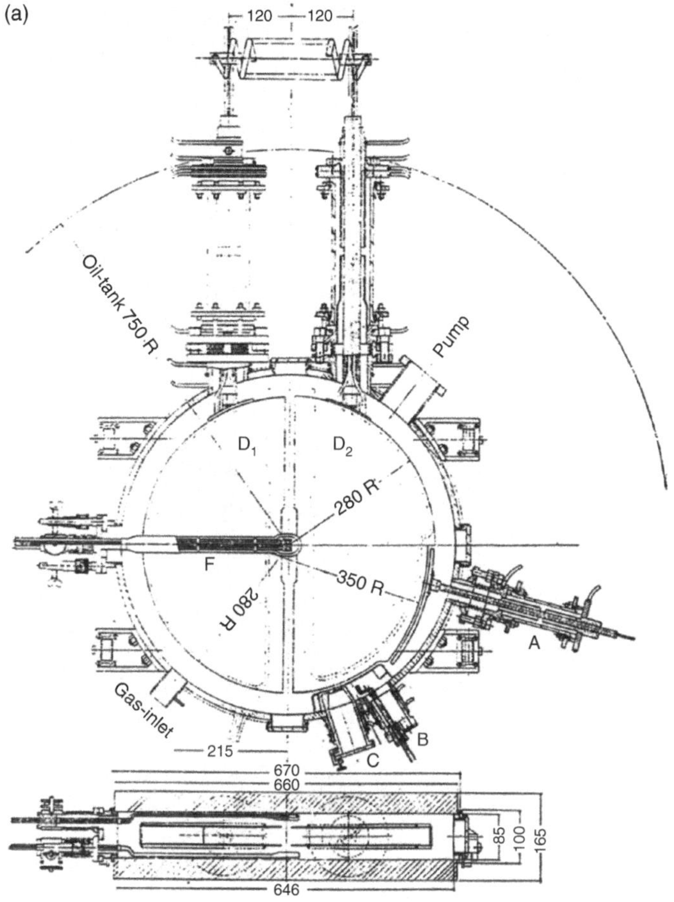

A. Deflector system; B. Target system; C. Observation port; D_1, D_2, dees; F. Filament system.

FIGURE 6.2 (a) Scale drawing of the vacuum chamber and accessories of Nishina's 26-inch cyclotron. (*Source:* Nishina, Yasaki, and Watanabe, *SP*, *34* [1938], 1658–1668.)

diffusion pumps (15 cm long and 7.5 cm in diameter) and an auxiliary air pump (for low vacuum) evacuated the chamber. The accelerating electrodes, or dees, consisted of copper sheeting 0.3 mm thick, 28 cm in radius, 3.5 cm in height, and 2 cm apart.

Two spiral tungsten filaments each of 1 mm in diameter generated the ions. Mounted in the central region between the dees, one above and the other below, they slowed in conducting a direct current of 45 A at about 7 V. An electrode set 5 mm away from the one dee and at a rectified potential of 50,000 V above or below it deflected the ions. The target chamber of copper strip made a Faraday cage divided into two sections inclined at an angle of about 120°. One section, made of copper plate, was used to measure currents; the other, of beryllium, offered a neutron-producing target. Most of the vacuum chamber parts were constructed in Riken's own workshop,

(b)

FIGURE 6.2 (b) Top: The photograph of the vacuum chamber and accessories, the cover plate being removed. Bottom: The photograph of the oscillator set (right) with cyclotron and pumps. (*Source:* Nishina, Yasaki and Watanabe, "The Installation of a Cyclotron," *SP*, *34* [1938], 1664 and 1666.)

where a large cloud chamber for Nishina's cosmic ray research previously had been built.[27]

The high frequency system consisted of an oscillator supplied by the Tokyo Electric Radio Company and a rectifier set.[28] "The SN167 oscillation tube, developed for short-waveradio communication," Hinokawa argued, "symbolized Japan's relative independence in this field and was the key to successful development of the small cyclotrons."[29] Althoughdirect current up to 12 kV could be supplied to the oscillator, experiments usually ran at about 6 to 7 kV and about 2.5 A. The rectifier set including a transformer was manufactured by the Nippon Denchi Kabushikikaisha. It could produce a wide variation of voltage "by changing the phase, through a phase shifter, of potential applied to the grids."

There were some significant differences between Nishina's 26-inch and Lawrence's 27-inch cyclotron.[30]Nishina's cyclotron produced 3 MeV deuteron beams

FIGURE 6.3 The front view of Nishina's 26-inch cyclotron. Notice two rabbits in glass boxes for biological experiments. (Courtesy of the Institute of Physical and Chemical Research.)

while Lawrence's up to 6.3 MeV; the shortest wavelength of the oscillating system of Nishina's was 30 m, longer than that of Lawrence's 20 m; and the pressure in the Tokyo vacuum chamber 10^{-3} mmHg as against 10^{-4} to 10^{-5} mmHg in Berkeley. Nishina made a few useful improvements, for example, Takahashi's automatic voltage regulator, which enabled Japanese researchers to avoid "tiresome hand control caused by the fluctuation of the thermal voltage of the generator" and to get a constant number of ions by synchronizing "the motion of ions and high frequency."[31]

The construction of the 26-inch cyclotron was completed on April 3, 1937. On that day Yasaki proudly reported to Lawrence the news of their first beam:

> Now I have the pleasure to inform you about the first operation of our cyclotron. When the pressure of deuterium became 5×10^{-4} mmHg, bright fluorescence was observed on the quartz target; and also enormous quantity of neutrons were observed when deuteron beams shot the beryllium target. At first the beam intensity was only 3×10^{-8} amperes, but after having adjusted the magnetic field as well as deflecting potential, it increased as much as one microampere. We are now endeavoring to get more current by adjusting high frequency voltage, because it seems to me that the beam intensity is exceedingly dependent upon this voltage. The magnetic field is 12,600 Gauss, and the energy of the deuterons is 2.6 MeV.[32]

At first the machine's operation was unstable and its beam was weak. Nishina acknowledged these difficulties in his letter to Lawrence on June 1: "Our cyclotron has been working since the beginning of April. Current is still small, 1 to 2 μA, which we are trying to increase." Six weeks later, on July 13, they had achieved only slight improvement: "Our cyclotron is working steadily at 2.9 MeV with 3 to 4 μA. We are trying to increase the current but have not succeeded yet."[33]

Sagane's return from Berkeley in February of 1938 boosted their efforts. He had learned much about the operation of Lawrence's cyclotron and even had published a full paper on radioactive isotopes of copper, zinc, gallium, and germanium using neutrons and deuterons produced by the Berkeley cyclotron.[34] Sagane,with the help of his assistant, Eizo Tajima, made several modifications in Nishina's cyclotron. They replaced the glass pipes that supported the dee arms with copper ones, and added safety boxes to the air-pumps and the heaters.[35] They alsosolved the vexing technical problem of how to shim the top pole of the magnet by inserting pieces of iron into "the small space between the cover of the vacuum chamber and the top pole of the magnet to correct the inhomogeneity of the magnetic field by the magnet."[36] Sagane's shimming technique made it possible forthe cyclotron's operators to control the intensity and stability of the ion beam. Elated, Nishina wrote to Lawrence, "Due to [Sagane's] shimming . . . our cyclotron has increased its current from about 15 to nearly 30 μA."[37] By the fall of1938, Nishina could report that the cyclotron could produce deuteron beams "of 20 to 50 μA at 3.0 to 2.6 MV . . . for some 30 hours without interruption."[38] The details of thecyclotron were published in a special number of Riken's *Scientific Papers of the Institute of Physical and Chemical Research* for 1938 issued to celebrate the sixtieth birthday of Riken's president, Masatoshi Okochi.

6.1.3 The Fruits

Nishina wrote in the 1938 report that the first work done with the 26-inch cyclotron concerned "the biological effects of neutron."[39] This was partly because the initial beam was notyet strong enough for nuclear research. However, Nishina recognized the promising future of radio-biology and closely monitored Lawrence's research on the subject.[40] As he wroteBohr:

> After you left Japan, we have been improving our cyclotron and we get at present 3 to 4 microamperes at 2.9 MV, which is limited by the saturation of our magnet. We are trying to increase the current but have not succeeded yet. Our medical colleagues have been studying the biological effects of neutrons and have obtained some interesting results. We sent a note to "Nature," a copy of which I am sending you by a separate post.[41]

Nishina here referred to Masanori Nakaidzumi, Koiti Murati, and Y. Yamamura's work on the biological effects of neutron rays from a beryllium target hit with 2.8 MeV deuterons from Riken's cyclotron (Figure 6.4a).[42] The result was submitted to *Nature* on July 13, 1937, almost a year before any results of nuclear physics research were published.

Nishina invited and encouraged biologists and medical doctors to use the cyclotron for their research. Biologists such as Murati, Hiromi Nakayama, Yoshihito Sinoto, and Duhei Sato as well as physicians such as Nakaidzumi, Taro Takemi, and Nobutane Mori accepted. They examined the effects of strong radioactivity on plant and animal physiology, especially on the genetic mutations, and also employed radioactive materials as tracers.[43] They unanimously praised Nishina for hisfarsightedness and generosity.[44]

(a)

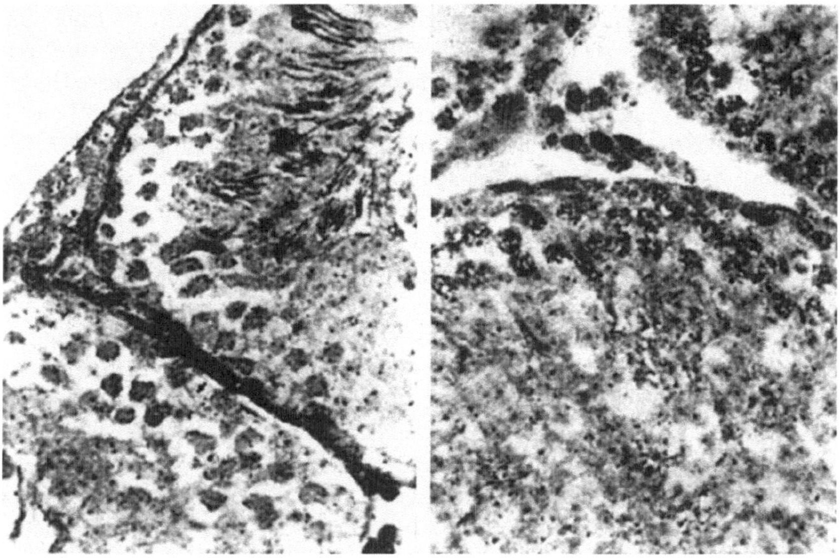

(b) TABLE I. *Summary of the identifications of fine products of uranium bombarded by fast neutrons.*

ELEMENT	HALF-PERIOD	ISOTOPE	RADIATION	MAXIMUM ENERGY (MEV)
Pd	26 min.			
	17 hr.	112	electrons	
Ag	7.5 day	111	electrons	0.8
	3.5 hr.	112	electrons	2.0
Cd	56 hr.	115		
	3 hr.	117		
	50 min.			
In	4.5 hr.	115*	γ-rays	0.31–0.35
			conversion electrons	0.28
	2 hr.	117	electrons	1.8

* Radioactive isomer of stable nucleus.

FIGURE 6.4 (a) The 26-inch cyclotron had been frequently employed for radioactive biology research. The left photo shows a normal testicle of a healthy mouse and the right one shows a testicle irradiated by "rays produced by bombarding a beryllium target with 2.8 MeV deuterons from the cyclotron." (*Source:* M. Nakaidzumi, K. Murati, and Y. Yamamura, *Nature*, *140* [1937], 359.) (b) When the 26-inch cyclotron could steadily produce strong beams, Nishina's nuclear research group began to produce several important results. (*Source:* Y. Nishina, T. Yasaki, K. Kimura, and M. Ikawa, *Physical Review*, *58* [1940], 661.)

Nishina too participated in biological research with scientists from Tokyo Imperial University. He collaborated with Nakayama in a study of the use of radioactive sodium, $_{11}Na^{24}$, as a tracer. The radioactive sodium was prepared by bombarding "3 MeV deuterons in the experimental chamber of the cyclotron."[45] Nishina, Takeo Iimori, Hideo Kubo, andNakayama used radioactive nitrogen (N^{13}, half-life 10.5 min), "obtained by bombarding amorphous carbon with 3 MeV deuterons from our cyclotron," to prove the possibility of the exchange between nitrogen gas and "some nitrogenous constituents of the plants."[46] Nishina joined with Sinoto and Sato inpublishing a series of papers on the effects of fast neutrons on several plants.[47] They placed various plant samples "in ametal box which was inserted in the irradiation chamber of the cyclotron" and exposed these samples for different time intervals to "neutron rays obtained by bombarding a beryllium target with 2.8 MeV deuterons." They carefully analyzed the neutron effects on external morphological features as well as on chromosomes. Daigoro Moriwaki and Nishina worked on the sex-linked mutations of *Drosophila* using fast neutron radiations, and confirmed that "male adults are more prominent in manifesting neutron-induced mutations than either pupae or larvae, just as was reported by Muller and Moor in the case of x-ray treatment."[48] Cooperation with biologists and medical menhelped Nishina not only to prove the versatility of the cyclotron but also to extend his research network outside Riken. These two factors were crucial in expanding his support from the Japanese science community for the construction of his 60-inch cyclotron.

Nuclear physics research using the 26-inch cyclotron started in the spring of 1938.[49] With the help fromShoji Kojima, Masao Ikawa, and Goro Miyamoto, Sagane led the first few experiments. His connection with Berkeley proved crucial, as some remarks from their papers indicated: "Germanium samples activated last year [1937] at Berkeley with 5.5 MeV deuteron beam from the cyclotron were brought back by one of the authors to Tokyo"; "A study of radioactivity produced by Y, Zr, and Mo started at the Radiation Laboratory, University of California, Berkeley, by one of the authors, has been continued by making chiefly neutron bombardments on these elements from the cyclotron in the Institute of Physical and Chemical Research, Tokyo"; "One of the authors takes pleasure in expressing his gratitude to Professor E. O. Lawrence . . . at the Radiation Laboratory, where many valuable techniques on the cyclotron were obtained and a few runs for preliminary tests on the present problem were made."[50]

Other members of the nuclear research group also used the 26-inch cyclotron to investigate artificial radioactivity by bombarding various elements with either fast deuterons or fast neutrons.[51] Yasaki and Watanabe bombarded oxygen gas andother oxides with 2 MeV deuterons. Yasaki wrote a detailed report on the fission products of uranium irradiated by fast neutrons that were "produced by bombarding metallic lithium with 3 MeV deuterons." Toshio Amaki and Asao Sugimoto investigated the β-ray spectra of Na^{24} and the energy levels of Mg^{24}. They also calculated the relative cross sections of (n, α) and (n, p) reactions caused by fast neutrons. Amaki, Sugimoto, and Iimori investigated the induced radioactivity of chromium bombarded with slow and fast neutrons as well as fast deuterons (3 MeV). Keizo Sinma and Fumio Yamasaki bombarded copper with "slow, fast neutrons from Li + D, and 3 MeV deuteron beams" and obtained the β-ray spectra of Cu^{62}, Cu^{64}, and Cu^{66}. The same team also got "strong β-active rhenium isotopes having half-lives of 16 and 90 hours, respectively" and measured "the relative capture cross sections for about 50 isotopes."

Nishina joined the nuclear physics research team. He worked with Yasaki and Hirohiko Ezoe, and secured the valuable cooperation of Kenjiro Kimura and Masao Ikawa of Tokyo Imperial University for the necessary chemical analysis. Their first experiment was on the artificial radioactivity induced in thorium by neutron bombardment: the researchers "obtained another thorium isotope which [they] identified with uranium Y [UY^{231}], the parent substance of protactinium, the thorium series thus being changed over to the actinium series."[52] The same researchers carriedout similar experiments with thorium in 1939 and discovered the fission products "of the following [radio-] active substances: Bi, Hg, Sb, Sn and Ag."[53]

In 1940, Nishina's nuclear research team began to bombard uranium with fast neutrons and published their results in the *Physical Review* and other journals.[54] They discoveredsilver, cadmium, indium, palladium, rhodium, and ruthenium isotopes among the fission products, and closely analyzed their decay curves. They achieved the same results almost simultaneously with their Western counterparts. For example, they discovered rhodium and ruthenium isotopes from uranium fission in 1941 almost at the same time as Emilio Segrè and Glenn T. Seaborg at Berkeley in 1941.[55] A paper in 1940 on the induced β-activity of uranium by fast neutrons also deserves special attention, since in it Nishina's team announced the discovery of U^{237}:

> A few grams of uranium oxide, U_3O_3, carefully purified and freed from its disintegration products were exposed to fast neutrons for more than fifty hours. After the exposure, a uranium fraction (U_3O_3) was separated and purified from all possible elements produced by fission as well as from its own disintegration products. The most care was given to the removal of lanthanum from the sample, the procedure taking as long as one day. The activity of the irradiated uranium was compared with that of a nonirradiated sample, in order to subtract the growing β-activity due to disintegration products of uranium. The difference thus obtained shows a 6.5-day period. This activity is probably due to U^{237} produced from U^{238} through loss of a neutron, as in the case of the production of UY from thorium. If this is the case, we have here a member of the missing radioactive family $4n + 1$.[56]

Nishina and his research team in the Nuclear Research Laboratory continued working on fission and uranium isotopes that would become an important part of a Japanese nuclear bomb project.

6.2 THE LARGE (60-inch) CYCLOTRON

6.2.1 First Try

Even before completing the first cyclotron in the spring of 1937, Nishina and his researchers were envisioning a larger one. As Yasaki indicated to Lawrence in January of 1936, the team considered their construction of the 26-inch cyclotron an exercise "for experience" needed for the construction of a larger cyclotron.[57] In the opening paragraph of his 1938 paper, "Installation of the Cyclotron," Nishina made this clear:

> One of the main items in its opening programme of the Nuclear Research Laboratory was the erection of a large cyclotron with an electromagnet weighing at least 100 tons. *As the preliminary to its execution*, we planned the installation of a medium sized one using an electromagnet of 23 tons.[58]

The Mitsui Ho-onkwai Foundation and the Japan Society's 10th subcommittee sponsored construction of the new, larger cyclotron.[59] The 10th subcommittee made an unprecedentedincrease in its budget in 1937 (127,850 yen) for the construction of large magnet.

The 60-inch cyclotron project had started in the summer of 1936. In July of 1936, he wrote to Lawrence:

> We are consequently trying to get a fund for the construction of a larger magnet of diameter 100 or 125 cm. In this connection I should like to know the cost of such a magnet in America and should be much obliged to you, if you would let Sagane know the suitable manufacturer in your country.[60]

Following Lawrence's advice, Nishina ordered a 60-inch electromagnet from Columbia Steel in San Francisco, a purchase that was brokered through the Mitsui Company.[61] However, theinexperience of the Mitsui representatives delayed the negotiations for this purchase despite Lawrence's intervention. Sagane returned to Berkeley in June of 1937 to accelerate the business. Meanwhile Nishina did not hesitate to ask Lawrence more detailed questions:

> I should like to know the necessary power for exciting the magnet. As I informed you by cable some time ago, we should like preferably to use 500 volts for this purpose, although it is not absolutely necessary. It might be necessary to install a new D. C. generator for it as the case may demand, and I should be much obliged to you, if you would let me know the necessary voltage and current for this purpose.
>
> I should like also to know the method of cooling of the magnet coils. If they are to be cooled by oil, oil tanks have to be made, and I should be much obliged to you, if you would let me know whether such tanks are included in our plan.[62]

The large electromagnet, coils and other materials that Nishina had ordered from the United States arrived in Japan in the spring of 1938.[63] The Ishikawashima Shipbuilding Company workedon the magnet for another three months before delivering it to Riken in June. Nishina happily wrote to Lawrence: "The machining of our magnet steel has nearly been finished and we have started erecting the magnet. All the coils and oil tanks have just arrived here and we are contemplating how to build them up."[64] The TokyoElectric Company provided the high-frequency system and Riken's workshop built most of the remaining parts. In February of 1939 preliminary test started. Completion was delayed for a few more months, however, because the war between China and Japan, which started in 1937, had created a serious shortage of parts. Sagane complained to Lawrence: "As we have chosen the same oscillator

(a)

FIGURE 6.5 (a) Nishina's 60-inch cyclotron. Nishina spent most his time reconstructing the machine to obtain stronger and more stable beams. (Courtesy of the Special Collections, NCSU Libraries.)

(b)

(c)

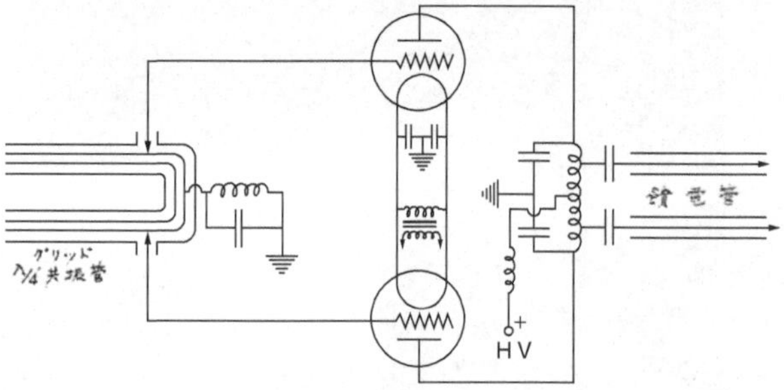

第 10 圖 發 振 器

FIGURE 6.5 (b) The photograph taken in front of the magnet for the 60-inch cyclotron. Nishina stood at the sixth from the left in the second row, Hantaro Nagaoka, seventh in the same row. (Courtesy of the Institute of the Physical and Chemical Research.) (c) Diagram of the oscillator for the 60-inch cyclotron. TW 530B oscillating tube (right) did not produce sufficiently strong output. (*Source:* K. Sinma, et al., *Kagaku Kenkyujo Hokoku, 27* [1951], 156–172 on 164.)

tubes with which the army is now constructing many broadcasting stations in China, all tubes produced by the company are taken away as soon as they are finished."[65] Meanwhile,Lawrence had finished his 60-inch cyclotron and had begun to operate it successfully.[66] He wrote Sagane in June of1939: "You can well imagine the thrill we have just had in seeing a deuteron beam extending out in the air for a distance of more than a meter and a half."[67]

Nishina's large cyclotron did not produce the anticipated strong beams. The electromagnet did not produce a strong enough field because the gap between the pole pieces was too wide; the ion chamber did not hold a vacuum rare enough;

and the oscillator, constructed on the same principle as that employed on the small cyclotron, did not deliver a frequency high enough.[68] Inaddition, Nishina and his collaborators did not consider changes that scale-up might require, such as the increase of particle mass at high speed and the importance of the minimum voltage.[69]Nishina's team struggled for almost a year to discover and correct the sources of the trouble. At first Nishina was optimistic. He wrote Rabi in February of 1940: "We shall soon be able to begin the operation of 60-inch cyclotron in our laboratory and hope to obtain various radioactive elements of sufficient intensities."[70] A few months later, he wrote Bohr that "Theoperation of our 200 ton cyclotron is not yet in right conditions, but I hope that will soon be brought in order."[71] By thesummer of 1940, however, Nishina and his colleagues had abandoned hope of being able to correct the problems themselves.

Desperate, on Jun 15, 1940, Nishina wrote Lawrence:

> During the coming summer vacation, two members of our laboratory, T. Yasaki and S. Watanabe, will be sent to America to visit some nuclear physics laboratories. I should be very much obliged to you, if you would kindly allow them to stay in your laboratory for about a month . . . .
>
> I have been thinking of coming myself to [the] Mecca of cyclotronists in order to have the pleasure of your personal acquaintance and also of learning various matters concerning the 60″ cyclotron, our edition, of which is not in right operating conditions yet. My health, however, does not allow me to leave here at present and I let my assistants come to you instead.[72]

To his surprise and joy, Lawrence accepted Nishina's request. World War II had already started in Europe and the relationship between Japan and the United States was deteriorating day by day. Nishina knew that it would be the last time he was likely to receive help from Lawrence. When Yasaki, Watanabe, and Takeo Imori arrived in San Francisco in August, Lawrence informed them that the president of the university would not allow Japanese guests to visit the Radiation Laboratory.[73] Yasaki and his colleaguesdid their best to acquire the information they needed but American physicists were far less cooperative than before. In 1943, Cooksey drafted a letter for Lawrence to M. Stanley Livingston that commented on that visit:

> When the Japs left this country [in the fall of 1940], they were making every effort to obtain from us the final details about our construction and methods. We have consistently refused them any information for approximately the last three years.[74]

The Japanese delegates did not leave empty handed, however. They received a rough blueprint of Berkeley's 60-inch cyclotron and some published papers, and also purchased a new Kinney pump (600 l/min).[75] Nishina expressed his gratitude to Lawrence:

> In view of the present international situation, I thank you from the bottom of my heart for what you have done for us in the cause of science and friendship . . . . From what I have heard I have decided to reconstruct various parts of our 60″ cyclotron according to your line of construction.[76]

6.2.2 Second Try

The Reconstruction started at the end of 1940 and was carried out solely by Japanese researchers without further assistance from Berkeley. For the task, Nishina mobilized Yasaki, Sagane, Sinma, Yamasaki, Tajima, Irie, and Sugimoto. They added iron discs of 1-inch radius to the poles to reduce the gap and increase the field; introduced a new oil tank with an automatic safety control box to cool the electromagnet's coils more efficiently; improved shimming; redid the vacuum chamber to overcome the minimum voltage of the dees more easily without the input of higher voltage; replaced the old pumps with the Kinney pump and a Hypervac; designed a new tungsten filaments for the ion source following Berkeley's design; and adopted the new resonance method to apply high voltage to the dees.[77]

Throughout this process, the cyclotron reconstruction team faced severe shortages in equipment, parts, and supplies. Even before the outbreak of the Pacific War, escalating economic pressures by the Western powers, culminating in a U.S. embargo on scrap iron and a de facto embargo on oil, made it difficult for researchers to import instruments and materials from the United States and elsewhere. The Japanese surprise attack on Pearl Harbor on December 7, 1941 severed vital Berkeley connection. The realities of the Pacific War interfered with shipping, and Japan's divergence of strategic materials to her war effort further reduced access to materials of all kinds, especially strategic materials. Serious shortages were felt in every sector of Japanese society, including at Riken. Nishina's team, for example, had difficulty locating suitable substitutes for the oil for the Kinney pump. They had to manufacture their own zinc disks to substitute for brass disks, and make rubber plates from crude rubber. Researchers carried on many of their researches in the street market and scrap yard.[78]

When, on February 11, 1943, preliminary test of the reconstructed cyclotron began, it failed to produce strong, steady 15 to 20 MeV deuteron beams its builders expected.[79] They workeddiligently for several more months to discover and fix the problems. On December 8, 1943, they first obtained 9 MeV proton beams, and, in February of the following year, at an intensity of 4.5 μA. The researchers then increased voltage between the dees and modified the cooling and high-frequency systems in order to accelerate deuterons. During the early summer of 1944, Nishina's team members were able to investigate relationships between the energy (MeV) and intensity (μA) of proton or deuteron beams. Thanks to the increase of voltage between the dees, they secured deuteron beams of 350–400 μA at 10 MeV and proton beams of 180 μA at 14 MeV.

From July of 1944, Sinma, Yamasaki, Tajima, Sugimoto, and a few others began using the 60-inch cyclotron to do research in physics.[80] Sinma worked on the fission of U^{235} byneutrons and the application of radioactive materials to luminous paint; Tajima, on the capture of slow neutrons in uranium; Sugimoto, on artificial polonium and absorption of neutrons in uranium; and Yamasaki on the proportion of U^{235} in natural uranium using neutrons. Since most of these studies had some relevance to the Japanese nuclear bomb project, their results remained largely unpublished. The only exception that clearly indicated the use of the 60-inch cyclotron was Sugimoto's paper on the α-activity of Bi bombarded by deuterons. "The author," Sugimoto wrote: "could investigate this problem with about 13 MeV deuteron beam

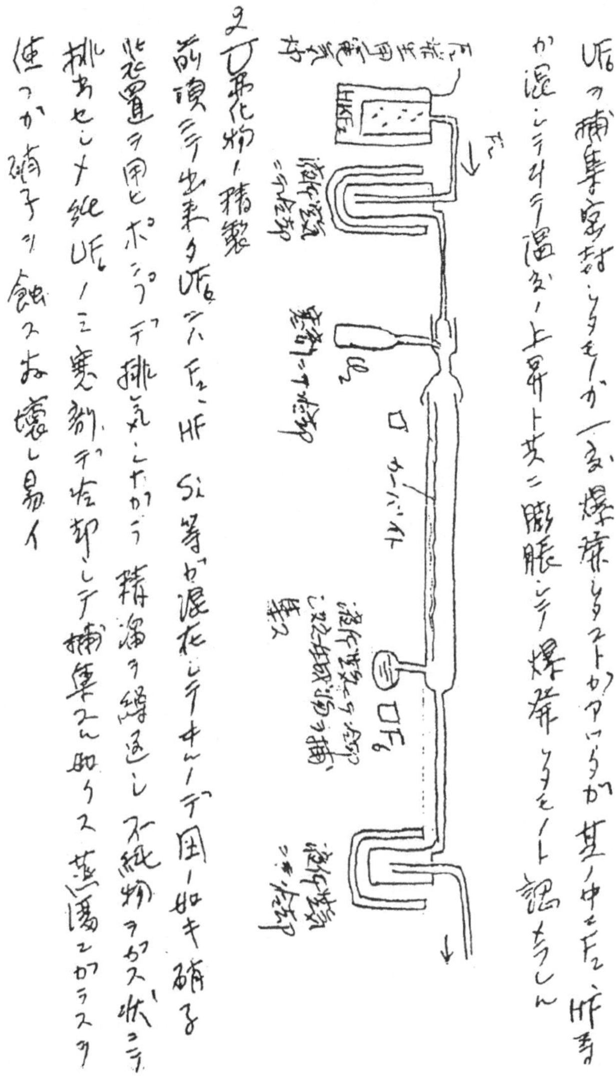

FIGURE 6.6 A diagram from the memo that recorded the dialog between Nishina and Army General Nobuuji. Japanese Army decided to choose gaseous discharge method to acquire necessary amount of U^{235} for the nuclear bomb. (Courtesy of the Institute of Physical and Chemical Research.)

from the 60-inch cyclotron in our laboratory and this paper is the report of the results."[81] Although not published until 1946,Sugimoto's research took place during the war. It suggests that much more work was done with the 60-inch cyclotron than is now known.

During World War II, with the support of the Japanese Army, Nishina headed a project aimed at developing nuclear bombs, the so-called Ni-Go Project (Figure 6.6).[82] Although the Ni-Goteam eventually adopted the thermal diffusion method to separate a tiny percentage of the U^{235} isotope from the much richer U^{238}, it also used the 60-inch cyclotron for the nuclear bomb project. Nishina insisted to Army Major General Nobuuji, who was responsible for directing the project, on the importance of the cyclotron in producing the necessary neutrons for nuclear bomb research.[83] Many historians and Japanesephysicists have argued that Nishina's involvement in and emphasis of the Army's nuclear bomb project was based primarily on his desire to save a greater number of young Japanese physicists from military conscription. Interestingly, in his few writings aimed at a general audience, Nishina exhorted the Japanese people and scientists in particular to wholeheartedly serve the war effort while, at the same time, emphasizing the importance of pure science. For example, in "The Reconstruction of the Greater East Asian Co-Prosperity Sphere and Pure Science," published in *Kagaku* in March of 1942, he wrote:

> Today the first and only goal of Japanese people is winning the war, and we, scientists, should devote our whole intellectual ability to serving for the war purpose . . . .
>
> It is not easy to reconstruct the Greater East Asian Co-Prosperity Sphere [the new Asia as envisioned by Japanese war planners, which would have included Japan, Eastern Siberia, China, Indo-China, the South Sea, and island groups in the Pacific] as well as winning the war . . . . [To achieve these two goals] science and technology have broad duty. Designing new weapons, developing industry, controlling of shortage or abundance of resources, all these activities demand the active involvement of scientists and engineers. The shortage of scientists and engineers therefore has been expected and the government has tried hard to increase the number of them.
>
> The shortage of both human and natural resources made people neglect pure science. However, we, scientists, should do our best to develop pure science while completing our duty mentioned above during the war. First, the advance of pure science will bring us the rapid development of technology as the history of science proves. For the one-hundred year construction of the Greater Asian Co-Prosperity Sphere after the war, the advance of pure science as well as technology is absolutely required.
>
> Second, the development of pure science should be made here in Japan for the whole Greater Asian Co-Prosperity Sphere, since the other parts of the Sphere are not ready to pursue it. It is also very difficult for each part of the Sphere to seek the development of pure science. Recently I had an opportunity to examine science and technology in Manchuria. Their primary goal is to utilize natural resources in Manchuria. They concentrated on applied research only and therefore have very few chances to pay enough attention to the development of pure science. It seems so natural for me and I had similar experience last year when I visited Taiwan. The situation may become very similar in other parts of the Greater Asian Co-Prosperity Sphere. So, Japan should take over the burden to develop pure science for the whole Sphere and other parts will use the results of it.
>
> Third, the War made it very difficult for Japan to import freely science from the West as it had done in the past. From now on, we should create our own science. However, we are far behind the West in number of scientists, budget for science, or natural resources. We should therefore double or triple our effort to compete with the West. When the war is over, and we open the box, if our science is far behind those of other

> [Western] countries, we scientists must take the blame and it will deter building the new order of the Greater Asian Co-Prosperity Sphere. Scientists then should be shamed in front of the military.[84]

The Japanese version of the Manhattan project was, however, much smaller, far less organized, and, by comparison, poorly financed. In the assessment of Kart T. Compton, who served in Japan as a member of the U.S. Scientific Intelligence Mission after the Japanese surrender, the traditional rivalry — even hatred — between Japan's Army and Navy was so serious that no systematic coordination had been made to form a single unit for the development of the nuclear bomb.[85] The Japanese Navy operatedits own nuclear bomb project, the F-Go Project, with physicists in Kyoto, but Nishina's team barely communicated with its naval counterpart. After the end of the war, Bunsaku Arakatsu, the builder of the Kyoto Imperial University cyclotron who had been involved in the F-Go project, told an American intelligence officer that nuclear research in Japan during the war "looked very well on paper, but really amounted to very little because nuclear physicists in different Japanese laboratories did not work in coordination."[86] A recent study by Keiko Nagase-Reimer,Walter E. Grunden, and Masakatsu Yamazaki showed that shortages of human and natural resources also were major obstacles to Japan's development of a nuclear bomb:

> Given the severe demands of the war on so many other areas of the national economy and science infrastructure, it is questionable whether Japan even had enough scientists, engineers, and technical personnel with sufficiently advanced training and experience available to dedicate solely to the effort to build a nuclear weapon from scratch, especially under wartime conditions .... [T]he number of principal scientists conducting research for the Navy's F-Go project appears not to have exceeded more than about twenty at any given time. The Army's Ni-Go project fared somewhat better, claiming some 30 principal researchers with an additional 110 scientists at Riken available for consultation and assistance.[87]

By the time Nishina began his nuclear bomb research in 1943, he already had lost a number of able researchers to other laboratories or to more urgent war-related projects such as the ill-fated radar and death ray projects.[88] Starting inmid-1944, constant air raids caused researchers to stop their work frequently to evacuate their research facility. Air raids on April 13 and 14 of 1945 destroyed two thirds of Riken's buildings, including the one that housed Nishina's small cyclotron. Although the small cyclotron was destroyed, to everyone's surprise, the large cyclotron survived, and Nishina spent most of his time operating it until the very end of the war. It was rumored that, after returning from his inspection of Hiroshima following the atomic bomb attack on that city on August 6, 1945, his first words were, "Is the cyclotron O.K.?"[89]

Nishina's beloved large cyclotron and remnants of small cyclotron, along with the other two Japanese cyclotrons located in Kyoto and Osaka, were all destroyed in November of 1945 by the order of the U.S. Joint Chiefs of Staff.[90] Photos published in *LifeMagazine* on December 24, 1945 (Figure 6.7), showed American soldiers dismantling the large cyclotron and dumping some of its parts into the sea,

FIGURE 6.7 Above left: American soldiers destroying Nishina's large cyclotron in November 1945. Above right: Nishina pleading American soldiers for his equipment. Below left: Dismantled magnet being dragged out from the building. Below right: The demolished cyclotron parts being dumped into sea-water. (Courtesy of the Special Collections, NCSU Libraries.)

"4000 ft deep so Japs could never retrieve them."[91] A photo caption proclaimed: "ProfessorNishina pleads for his equipment. 'This is ten years of my life,' he said. 'It has nothing to do with bombs.' His wife and secretary wept quietly." Many American scientists severely criticized the action, and the Secretary of War later claimed that it was a mistake. Nevertheless, it was quite clear that the American military destroyed the Japanese cyclotrons without hesitation or consultation with

American scientists. Thomas C. Smith, a retired history professor at the University of California, Berkeley, who had served as an interpreter in 1945, remembered:

> It was a chilly morning in late October or early November when I checked in at my group's office, where the Colonel seemed to be waiting for me. He asked if I knew how to get to Kyoto University .... I assured him that I had been to the University several times and could get there without difficulty. He then told me that he wanted me to act as guide and interpreter for two officers from the Navy Department in Washington who had arrived the day before on an important mission. They were to oversee the dismantling of the cyclotron known to be at Kyoto University, which a company of Army engineers were standing by to cut up with blow torches preparatory to the Navy dumping them in a deep sea trench off Japan where they could never be recovered ....
>
> I have often wondered since why the destruction had to be completed the very day we heard about it, though I did not question that judgment at the time. Now I suspect it was to get the deed done before news of the event was known in the United States, where American physicists may well have opposed its destruction. I also wondered why one or both of the Japanese American interpreters in my section were not used ahead of me, since their Japanese was considerably better than mine. I doubt that the Colonel himself made the decision to exclude them because, as recall, he had been with the two men for a considerable time during the war and seemed genuinely fond of them. A more likely possibility is that the few men in Washington who made the decision to destroy the cyclotron specified that the interpreter be a Caucasian, thinking that such a person was less likely to breach security about what had happened than a Japanese American. In this regard it is worth mentioning that the Navy went to the immense expense of sending several hundred people like myself through the Naval Japanese Language School at Boulder, Colorado, rather than recruit Japanese Americans for the job, as the Army did.[92]

6.3 THE LEGACY OF TWO CYCLOTRONS

The nuclear research group in the Nishina Laboratory shared several features in common with the cosmic ray research group. First, the two groups consisted of men with diverse educational backgrounds. Although physicists from Tokyo Imperial University formed the majority, many researchers came from other departments of that university and from other universities. Second, the nuclear research group lacked a charismatic leader like Tomonaga of the theory group. Nishina was personally involved in the construction and operation of two cyclotrons and also produced several papers with young collaborators. Sagane might have become the independent leader in the nuclear research group. He had been trained in Berkeley and Cambridge, contributed greatly to the construction and operation of the 26-inch cyclotron, and he was the first Japanese to publish a paper on nuclear physics using the small cyclotron. However, Sagane left the Nishina Laboratory in 1939 to become a physics professor at Tokyo Imperial University, after which he no longer contributed greatly to the construction and operation of the large cyclotron. Sagane's sudden departure is rumored to have been precipitated by Nishina's disappointment with Sagane's contributions to the rebuilding of the 60-inch cyclotron. Nishina then transferred his confidence to Yasaki. Third, like the cosmic ray group, the nuclear research group failed to produce spectacular results with either of its two impressive machines. Although many of their findings were good enough to be published in *Nature* or the *Physical Review*, none were good enough to be remembered in physics textbooks.

The construction and operation of the two cyclotrons in the Nishina Laboratory, however, delivered important but different contributions to the development of physics in Japan. Above of all, Nishina's cyclotron project indicates the emergence of big science in Japan during the 1930s and 1940s. Big machines, huge sums of money from the Japanese government and industry, the construction of special instruments by major Japanese companies, and the mobilization and interdisciplinary use of hundreds of scientists and engineers for a single project.[93] Between 1936and 1945, the project bred a new generation of Japanese physicists and engineers with expert knowledge of the cyclotron. Their knowledge and experience would greatly contribute to the rapid growth of postwar physics in Japan. In an interview, Nambu recounted an interesting episode.[94] After the war, Nambu andother physics students at Tokyo Imperial University took shelter in a small lecture room. The room was crowded and everything had been destroyed: they had nothing to do but sit and talk. There Nambu met a senior physicist who had participated in the large cyclotron project, and that man told Nambu almost all that he knew about the cyclotron. Although Nishina's cyclotrons had been destroyed, the knowledge and experience acquired in building them survived.

On the other hand, the construction and operation of the two cyclotrons demonstrated the limitations of Japanese science and technology in the early twentieth century. In 1937, when Nishina asked a Japanese naval dockyard for a price estimate for constructing the 60-inch electromagnet, the estimate was almost double that provided by the American company, Columbia Steel, which had been recommended to Nishina by Lawrence and from which Nishina eventually obtained the needed electromagnet.[95] Construction of the large cyclotron made Japan's technological backwardness more apparent. The termination of help and advice from Berkeley in late 1940 and the stoppage of shipping of parts and supplies from the United States to Japan in 1941 seriously compromised the ability of the Japanese researchers to rebuild their large cyclotron. Japanese companies failed to manufacture the critical instruments researchers needed and to provide parts or part repairs on time. Although Japan's war effort originally had made it possible for Nishina to find funding for his 60-inch cyclotron project, ironically this project was now doomed, according to Hinokawa, by the military's huge demand for weaponry:

> The development of a large cyclotron required the development of a new oscillating tube that had greater output than existing high-power short-wave oscillating tubes.
>
> But the [Japanese] military demand that had spurred the rapid development of Japan's electronics industry in the 1930s had, in the final stages of World War II, turned the entire industry into "factory-wide battlefields on which to serve the country with radio waves," and it was impossible to develop new technology in divisions having no direct connections to weaponry. This clearly illustrates the way the progress of technology is hindered when it is turned solely to meeting military demand, consequently placing limitations on scientific progress. The abnormal situation this created also took its toll on researchers like Imaoka [the engineer at the Tokyo Electric Company who developed

> the SN-167 tube], who was compelled to direct plant operations to meet the military's production demands; he became neurotic and died on March 14, 1944.
>
> From the standpoint of experiments using the cyclotron, insufficient oscillating output meant, first of all, that the accelerating electrodes could not be loaded with high voltage at high frequency, so the beam produced would inevitably have had weak intensity. Experiments using a weak beam would have necessitated an internal target. And, in fact, Shinma and his associates later wrote that, based on the results achieved:
>
> > "... we judged that it was adequate for experiments on an interpolated [*sic*] target, but that the rise in Dee voltage was not yet sufficient to draw the beam outside. So, we finally decided to experiment with an interpolated [*sic*] target. At this point, we decided to suspend overall adjustments and begin using the large cyclotron for our personal experiments from around July 10."
>
> That was in 1944, seven years after the project to develop a large cyclotron had been launched.[96]

The cyclotron projects also highlighted Japan's shortage of human resources in science and technology. Although Nishina was an excellent physicist and organizer, he was not the equal of Lawrence in cyclotron construction and operation. He also lacked the array of talented assistants and colleagues who so ably assisted Lawrence, such as M. Stanley Livingston, Robert Serber, Donald Cooksey, Luis W. Alvarez, John J. Livingood, Glenn T. Seaborg, and Edwin McMillan, three of whom would become Nobel Laureates in physics (Alvarez in 1968) and chemistry (Seaborg and McMillan in 1951).[97] No foreign experts like Fermi andSegrè visited Riken to offer Nishina advice nor could he find in Japan the equal of theorists like Robert Oppenheimer to consult with him or critique his work. The relatively few Japanese theoretical physicists, including Yukawa, Tomonaga, Sakata, and Taketani, seldom paid attention to cyclotrons. In short, Berkeley sat at the center of a worldwide network, while Nishina's nuclear physics group was increasingly isolated.

Why did Nishina skip the medium-size cyclotron (30–40-inch with 8–12 MeV)? Lawrence and his Berkeley researchers had first modified one of small cyclotrons to a medium size before constructing a large 60-inch model. In doing so, they accumulated expertise that helped them to avoid many fatal errors when designing the large machine. Why was Nishina in such a hurry? Was he overconfident? Had he overestimated his strength? Nishina became overly optimistic after completing the small cyclotron. In the summer of 1937, he wrote to Hevesy:

> We are going to install an electromagnet of about 220 tons, by means of which we ought to be able to produce a beam of 25 MV, if it works successfully. Professor Lawrence of California University has very kindly ordered the magnet for us and *we hope to complete it some time next year [1938]*.[98]

It is revealing that almost no work with the small cyclotron was reported in Riken's *Scientific Report of the Institute of Physical and Chemical Research* after the fall of 1941. Nishina had put most of the physicists at the Nuclear Research Laboratory, but especially the experimentalists, on the reconstruction of the 60-inch cyclotron.

In retrospect, it seems obvious that these researchers could have produced more important scientific results using the small cyclotron, as the research done from 1938 to 1941 proves. Was Nishina's decision to mobilize his researchers for the reconstruction a great mistake? How did Nishina's subordinates appraise this decision? Tomonaga answered this question in his eulogy of Nishina:

> At times, Dr. Nishina was too far-sighted. Given the situation of our country, there were times when his plans would probably have worked if he had been a bit less ambitious. Since we had a small cyclotron, we could have used it to carry out various research projects which, though small in scale, would have been useful. However, Dr. Nishina was not one to be contented with small successes . . . .
>
> In fact, establishing the foundation for research was more important to him than the research itself. This task of laying the groundwork was so difficult that he was not able to do anything more. The problem lay in the situation of our country at that time. In order to build a large cyclotron, the entire staff of the Atomic [*sic*] Nuclear Research Laboratory of the Institute of Physical and Chemical Research had to shoulder many tasks, including fund raising, collection of materials, negotiations with contractors, and some things which engineers would naturally be expected to handle. This is why we could not make full use of the already completed, operational small cyclotron. It was just like the proverb says: "Poor men have no leisure."[99]

Tomonaga's message, though circumspect, seems clear: most Japanese physicists of his day did not agree with Nishina's decision to concentrate on the 60-inch cyclotron.

How successful were Nishina's cyclotrons? Because of the long delay in the completion of the 60-inch cyclotron, the connection between the large cyclotron and the Japanese nuclear bomb project, the destruction of Nishina's two cyclotrons by the American military, and the general reluctance of Japanese scientists to discuss their wartime activities, many historians and Japanese physicists (especially those who actually participated in the project) still describe Nishina's cyclotrons as at best a "half success." Rather, they should be considered great successes. Nishina's small cyclotron, built in 1937, was the first working cyclotron outside the United States.[100] Itallowed Japanese researchers to produce many top-quality papers on nuclear physics and radioactive biology. If the 60-inch cyclotron had been completed on schedule in 1939 according to the original schedule, it might have been the largest or at least the second largest in the world, ranking just after Lawrence's 60-inch cyclotron. Nishina and his team managed to complete their large cyclotron under very adverse wartime conditions. Once again, although much delayed, it was the largest operating cyclotron outside the United States before 1945.[101]

In many respects, Nishina's two cyclotrons, especially the larger one, should be considered as successful examples of reverse engineering, a hallmark of Japanese technology in the interwar period. To engineer in reverse means to replicate every part of a machine without either detailed blueprints or licensed know-how. This requires "an understanding of underlying scientific principles, and an ability to adapt foreign models to the needs of the domestic market."[102] Historians of science therefore should put more emphasis on how Japanese scientists overcame their difficulties and successfully adapted imported knowledge to their construction of cyclotrons rather than emphasize the influence of Western science and technology (in this case, Lawrence at Berkeley) on Japanese work.

NOTES

1 Asahi Shimbun, *1937 Asahi Nenkan*, (Tokyo: Asahi Shimbunsha, 1938), p. 353.
2 Caption from "Cyclotron Smashing: American Soldiers Demolish and Sink Precious Jap Scientific Equipment," *Life*, *19:26* (December 24, 1945), 26–27 on 27.
3 H. Tamaki and H. Ezawa (eds.), *Nishina Yoshio*.
4 Morris-Suzuki, *Technological Transformation of Japan*, Chapter 6.
5 Hirosige, "Social Conditions for Prewar Japanese Research in Nuclear Physics," p. 218.
6 Asahi Shimbunsha, *1937 Asahi Nenkan*, p. 353.
7 Oh, *Nishina Yoshio and Modern Physics in Japan*, pp. 179–181. See also T. Hirosige, "Social Conditions for Prewar Japanese Research in Nuclear Physics," p. 214.
8 Shizue Hinokawa, "Cyclotron Development at the Institute of Physical and Chemical Research in the 1930s," *Tokyo Institute of Technology Studies in Science, Technology and Culture*, *4* (2001), 14–37 on 26–27.
9 Ibid., 17.
10 Mitsui Ho-onkwai Foundation, *Prospectus of the Mitsui Ho-onkwai Foundation* (March, 1944), p. 81.
11 Y. Nishina, "Recent Development of High Speed Accelerators [in Japanese]," *Kagaku*, *9* (1939), 441–445; Y. Nishina, "1939 Nobel Prize Winner in Physics, Ernest O. Lawrence [in Japanese]," *Kagaku Chishiki* (January, 1940), 110–114.
12 "Nishina Laboratory" in "Summary of the Past Activities of the Institute," *SP*, *34* (1938), 1842–1849 on 1847.
13 "Nishikawa Laboratory," in "Summary of the Past Activities of the Institute," 1837–1841 on 1840.
14 "Nishina Laboratory," 1847.
15 For the list of major participants of the Nuclear Research Laboratory, see Oh, *Nishina Yoshio and the Modern Physics in Japan*, p. 196.
16 Ibid., p. 198.
17 Y. Nishina to N. Bohr (August 3, 1936) in *Y. Nishina's Correspondence with N. Bohr and Copenhageners, 1928–1949*, pp. 57–58.
18 Y. Sugiura and O. Minakawa, "On the Neutron Groups," *SP*, *34* (1938), 1299–1307; K. Shinohara and M. Hatoyama, "Pair Production by Gamma Rays from Fluorine Bombarded with Protons," *SP*, *38* (1941), 253–262 and 326–327; M. Hatoyama, Ziro Yuhara, and T. Maeyama, "Pair Production by Gamma Rays from Fluorine Bombarded with Protons. III," *SP*, *39* (1941), 1–7; K. Shinohara, M. Hatoyama, and Z. Yuhara, "Pair Production by Gamma Rays from Fluorine Bombardment with Protons. IV," *SP*, *39* (1941), 8–13.
19 Y. Nishina, T. Yasaki, and S. Watanabe, "The Installation of a Cyclotron," *SP*, *34* (1938), 1658–1668.
20 Nishina's team specifically mentioned following papers: Ernest O. Lawrence and N. E. Edlefsen, "On the Production of High Speed Protons," *Science*, *72* (1930), 376–377; E. O. Lawrence and M. Stanley Livingston, "The Production of High Speed Light Ions without the Use of High Voltage," *Physical Review*, *40* (1932), 19–35; E. O. Lawrence and M. Stanley Livingston, "The Multiple Acceleration of Ions to Very High Speeds," *Physical Review*, *45* (1934), 608–612; M. Stanley Livingston, "The Magnetic Resonance

Accelerator," *Review of Scientific Instruments*, *7* (1936), 55–68; E. O. Lawrence and Donald Cooksey, "On the Apparatus for the Multiple Acceleration of Light Ions to High Speeds," *Physical Review*, *50* (1936), 1131–1140.

21 Y. Nishina to E. Lawrence (April 21, 1932), Nishina Archive MSS 480. This is a reply to Lawrence's letter of December 5, 1931 (MSS 479).

22 John L. Heilbron and Robert W. Seidel, *Lawrence and his Laboratory: A History of the Lawrence Berkeley Laboratory*, *Vol. 1* (Berkeley: University of California Press, 1989), p. 318. In his letter to Tameichi Yasaki, Lawrence said, "I notice in the photograph that you have some holed drilled in the bottom plate of the vacuum tank in the same way as we had last year. I hasten to suggest to you that you plug these holes up, as we found that such holes in the top or bottom plate produce asymmetries in the magnet field which are almost impossible to correct by shims." [E. Lawrence to T. Yasaki (December 24, 1936), EOL, 9:43].

23 T. Yasaki to D. Cooksey (September 11, 1936), EOL, 9:43.

24 Y. Nishina to E. Lawrence (July 30, 1936 and February 21, 1938), EOL, 9:48.

25 Nishina, Yasaki, and Watanabe, "The Installation of a Cyclotron," 1659–1661.

26 Ibid., 1661.

27 For Riken's workshop during the 1930s, see "The Development of Riken [in Japanese]," *Shizen* (December, 1978), 15; "The Roundtable Discussion: The Role of the Workshop to Raise the Level of Research [in Japanese]," *Shizen* (December, 1978), 58–67.

28 Nishina, Yasaki, and Watanabe, "The Installation of a Cyclotron," 1665–1666.

29 Hinokawa, "Cyclotron Development at the IPCR," 19.

30 This comparison is based on Nishina's 1938 paper ("The Installation of a Cyclotron") and Lawrence and Livingston's 1934 ("The Multiple Acceleration of Ion to very High Speeds") and Lawrence and Cooksey's 1936 ("On the Apparatus for the Multiple Acceleration of Light Ions to High Speeds") papers.

31 Nishina, Yasaki, and Watanabe, "The Installation of a Cyclotron," 1661 and Hinokawa, "Cyclotron Development at the IPCR," 19–20.

32 T. Yasaki to E. O. Lawrence (April 3, 1937), EOL, 9:43.

33 Y. Nishina to E. O. Lawrence (June 1, 1937 and July 13, 1937), EOL, 9:38.

34 Sagane read the summary of the research in the 217th regular meeting of the American Physical Society in December of 1937 [*Physical Review*, *53* (1938), 211], and submitted the full paper a year later [R. Sagane, "Radioactive Isotopes of Cu, Zn, Ga and Ge," *Physical Review*, *55* (1939), 31–38]. See also W. Y. Chang, M. Goldhaber, and R. Sagane, "Absorption of γ-Rays Measured by Their Photo-Effect in Beryllium," *Nature*, *139* (1937), 962–963. This last work was done while he worked at the Cavendish Laboratory.

35 Eizo Tajima, "The Cyclotron in Riken [in Japanese]," Tamaki and Ezawa (eds.), *Nishina Yoshio*, pp. 119–127 on pp. 119–120. These glass (pyrex) tubes were employed in the Berkeley's 27-inch cyclotron "whose failure gave so much trouble with the 27-inch cyclotron" (Heilbron and Seidel, *Lawrence and His Laboratory*, p. 292).

36 E. O. Lawrence and M. Livingston, "Multiple Acceleration," 609.

37 Y. Nishina to E. O. Lawrence (February 21, 1938), EOL, 9:38.

38 Nishina, Yasaki, and Watanabe, "The Installation of a Cyclotron," 1667.

39 "Nishina Laboratory," 1849.

40 Letters from Lawrence to both Yasaki and Nishina as well as those from Sagane to Nishina were major sources of information: E. O. Lawrence to T. Yasaki (March 21, 1936, August 28, 1936, and December 24, 1936), EOL, 9:43; E. O. Lawrence to Y. Nishina (July 2, 1937), EOL, 9:38.

41 Y. Nishina to N. Bohr (August 28, 1937) in *Supplement to the Publications*, pp. 21–22.

42 M. Nakaidzumi, K. Murati, and Y. Yamamura, "Biological Effects of the Rays Produced by a Cyclotron," *Nature*, *140* (1937), 359.
43 Few examples are, M. Nakaidzumi and K. Murati, "Biological Effects of the Rays Produced by a Cyclotron," *SP*, *34* (1938), 357–361; M. Nakaidzumi and K. Murati, "Effects of Be-D Radiations upon *Vicia faba*," *Nature*, *142* (1938), 534–535.
44 See Hiromi Nakayama, "Tracer and Plant Physiology [in Japanese]" and Daigoro Moriwaki, "Nishina Sensei and Radiation Biology [in Japanese]" in Tamaki and Ezawa (ed.), *Nishina Yoshio*, pp. 135–153 and pp. 154–163 each.
45 Y. Nishina and H. Nakayama, "On the Absorption and the Translocation of Sodium in the Plant," *SP*, *34* (1938), 1635–1642 on 1637.
46 Y. Nishina, T. Iimori, H. Kubo, and H. Nakayama, "The Exchange Reaction between Gaseous and Combined Nitrogen," *Journal of Chemical Physics*, *9* (1941), 571–572.
47 Y. Nishina, Y. Sinoto, and D. Sato, "Effects of Fast Neutrons upon Plants, I. Abnormalities in *Fagopyrum* and *Cannabis* [in Japanese]," *Bulletin IPCR*, *18* (1939), 721–734; Y. Nishina, Y. Sinoto, and D. Sato, "Effects of Fast Neutrons upon Plants, II. Abnormal Behavior of Mitosis in *Vicia faba*," *Cytologia*, *10* (1940), 406–421; Y. Nishina, Y. Sinoto, and D. Sato, "Effects of Fast Neutrons upon Plants, III. Cytological Observations on the Abnormal Forms of *Fagoyrum* and *Cannabis*," *Cytologia*, *10* (1940), 458–466; Y. Nishina, Y. Sinoto, and D. Sato, "Effects of Fast Neutrons upon Plants, IV. Cytoplasmic changes in *Spirogyra*," *Cytologia*, *11* (1940), 311–318. The quotation comes from the part II of the series (p. 406).
48 Y. Nishina and D. Moriwaki, "Sex-Linked Mutations of *Drosophila melanogaster* Induced by Neutron Radiations," *SP*, *36* (1939), 419–425 on 422. See also Y. Nishina and D. Moriwaki, "Sex-Linked Mutations of *Drosophila melanogaster* Induced by Neutron Radiations from a Cyclotron II," *SP*, *38* (1941), 371–376.
49 For this line of research, see N. Saito, "Nishina Yoshio and Isotopes [in Japanese]," in Tamaki and Ezawa (eds.), *Nishina Yoshio*, pp. 128–134.
50 R. Sagane, S. Kojima, and M. Ikawa, "Radioactive as Isotopes," *Physical Review*, *54* (1938), 149–150 on 149; R. Sagane, S. Kojima, G. Miyamoto, and M. Ikawa, "Preliminary Report on the Radioactivity Produced in Y, Zr, and Mo," *Physical Review*, *54* (1938), 542–543 on 542; R. Sagane, S. Kojima, G. Miyamoto, and M. Ikawa, "Neutron Induced Radioactivity in Columbium," *Physical Review*, *54* (1938), 970.
51 T. Yasaki and S. Watanabe, "Deuteron-Induced Radioactivity in Oxygen," *Nature*, *141* (1938), 787; T. Yasaki, "Fission Products and Induced β-Ray Radioactivity of Uranium by Fast Neutrons," *SP*, *37* (1940) 457–472; T. Amaki and A. Sugimoto, "Beta-Ray Spectra of Radioactive Antimony and Sodium," *SP*, *34* (1938), 1650–1657; A. Sugimoto, "Energy Levels of the ^{24}Mg Nucleus," *Nature*, *142* (1938), 754–755; T. Amaki and A. Sugimoto, "On the Relative Cross Sections of the (n, α) and (n, p) Reactions produced by Fast Neutrons," *SP*, *38* (1941), 377–381; A. Sugimoto, "Note on the Relative Cross Sections of the (n–p) and (n–α) Reactions Produced by Fast Neutrons," *SP*, *39* (1941), 277; T. Amaki, T. Iimori, and A. Sugimoto, "Artificial Radioactivity of Chromium," *SP*, *37* (1940), 395–398; K. Sinma and F. Yamasaki, "Beta-Ray Spectra of Cu^{62}, Cu^{64} and Cu^{66}," *SP*, *35* (1938), 16–23; F. Yamasaki and K. Sinma, "Beta-Radioactivities of Rhenium," *SP*, *37* (1939), 10–16; K. Sinma and F. Yamasaki, "Capture Cross Sections for Slow Neutrons," *SP*, *38* (1941), 167–173.
52 Y. Nishina, T. Yasaki, K. Kimura, and M. Ikawa, "Artificial Production of Uranium Y from Thorium," *Nature*, *142* (1938), 874.
53 Y. Nishina, T. Yasaki, H. Ezoe, K. Kimura, and M. Ikawa, "Fission of Thorium by Neutrons," *Nature*, *144* (1939), 547–548 on 547. In the paper, they lamented the missing opportunity of the discovery of fission process, saying that "Chemical properties

suggested that either 'transuranic' or elements of lower atomic number than bismuth were involved, but both these alternatives were difficult to accept at that time."

54 Y. Nishina, T. Yasaki, H. Ezoe, K. Kimura, and M. Ikawa, "Induced β-Activity of Uranium by Fast Neutrons," *Physical Review*, *57* (1940), 1182; Y. Nishina, T. Yasaki, H. Ezoe, K. Kimura, and M. Ikawa, "Fission Products of Uranium Produced by Fast Neutrons," *Nature*, *146* (1940), 24; Y. Nishina, T. Yasaki, K. Kimura, and M. Ikawa, "Fission Products of Uranium by Fast Neutrons," *Physical Review*, *58* (1940), 660–661; Y. Nishina, T. Yasaki, K. Kimura, and M. Ikawa, "Fission Products of Uranium by Fast Neutrons," *Physical Review*, *59* (1941), 323–324; Y. Nishina, T. Yasaki, K. Kimura, and M. Ikawa, "Fission Products of Uranium by Fast Neutrons," *Physical Review*, *59* (1941), 677; Y. Nishina, K. Kimura, T. Yasaki, and M. Ikawa, "Einige Spaltproduckte aus der Bestrahlung des Urans mit schnellen Neutronen," *Zeitschrift für Physik*, *119* (1942), 195–200.

55 Nishina, Yasaki, Kimura, and Ikawa, "Fission Products of Uranium by Fast Neutrons," 677. See Emilio Segrè and Glenn T. Seaborg, "Fission Products of Uranium and Thorium produced by High Energy Neutrons," *Physical Review*, *59* (1941), 212–213.

56 Nishina, Yasaki, Ezoe, Kimura, and Ikawa, "Induced β-Activity of Uranium by Fast Neutrons," 1182.

57 T. Yasaki to E. O. Lawrence (January 31, 1936), EOL, 9:43.

58 Nishina, Yasaki, and Watanabe, "The Installation of a Cyclotron," 1658. Italics are added.

59 Oh, *Nishina Yoshio and the Modern Physics in Japan*, pp. 209–210, Hirosige, "Social Conditions for Prewar Japanese Research in Nuclear Physics," pp. 217–218, and Hinokawa, "Cyclotron Development at the IPCR," 17 and 25–28.

60 Y. Nishina to E. O. Lawrence (July 30, 1936), EOL, 9:38.

61 For the purchase of the magnet from Columbia Steel, see EOL, 9:37.

62 Y. Nishina to E. O. Lawrence (June 1, 1937), EOL, 9:38.

63 For the details of the construction of the large cyclotron, see K. Sinma, et al., "Report of the Construction of the 60-inch Cyclotron [in Japanese]," *Kagaku Kenkyujo Hokoku,* 27 (1951), 156–172. Unlike the small, 26-inch cyclotron, Nishina did not publish any detailed reports about the large cyclotron.

64 Y. Nishina to E. O. Lawrence (May 2, 1938), EOL, 9:38.

65 R. Sagane to E. O. Lawrence (July 24, 1939), EOL, 9:39.

66 E. O. Lawrence, et al., "Initial Performance of the 60-inch Cyclotron of the William H. Crocker Radiation Laboratory, University of California," *Physical Review*, *56* (1939), 124.

67 E. O. Lawrence to R. Sagane (June 16, 1939), EOL, 9:39.

68 Sinma, "Report of the Construction of the 60-inch Cyclotron," 156.

69 Ibid., 163; Tajima, "The Cyclotron at Riken," 122–123; Oh, *Nishina Yoshio and the Modern Physics in Japan*, p. 212.

70 Y. Nishina to Isidor I. Rabi (February 26, 1940), Box 6, Folder 4 (National Archive, I. I. Rabi Collection).

71 Y. Nishina to N. Bohr (June 15, 1940) in *Supplement to the Publications*, p. 26.

72 Y. Nishina to E. O. Lawrence (June 15, 1940), EOL, 9:38.

73 E. O. Lawrence to Y. Nishina (August 22, 1940), EOL, 9:38.

74 Draft letter from Lawrence to M. Stanley Livingston written by Cooksey (November 18, 1943), EOL, 12:12.

75 Tajima, "The Cyclotron at Riken," 123.

76 Y. Nishina to E. O. Lawrence (November 28, 1940), EOL, 9:38.

77 Sinma, et al., "Report of the Construction of the 60-inch Cyclotron," 158–169.

78 Oh, *Nishina Yoshio and Modern Physics in Japan*, pp. 212 and 218–219.

79 For the test of the large cyclotron, see Sinma, et al., "Report of the Construction of the 60-inch Cyclotron," 169–170, and Tajima, "The Cyclotron at Riken," 123–124.
80 Sinma, et al., "Report of the Construction of the 60-inch Cyclotron," 170.
81 A. Sugimoto, "On the α-Activity of Bi Bombarded with Deuterons," *SP, 42* (1946), 152–159.
82 There are very few reliable analyses about the Japanese nuclear bomb project during the World War II. See Pacific War Research Society, *The Day Man Lost: Hiroshima, 6 August 1945* (Palo Alto: Kodansha International, 1972); Deborah Shapley, "Nuclear Weapons History: Japan's Wartime Bomb Project Revealed," *Science, 199* (1978), 152–157; Morris F. Low, "Japan's Secret War? Instant Scientific Manpower and Japan's World War II Atomic Bomb Project," *Annals of Science, 47* (1990), 347–360; John W. Dower, *Japan in War and Peace: Selected Essays* (New York: New Press, 1993), Chapter 3: 'NJ' and 'F': Japan's Wartime Bomb Research; Robert Wilcox, *Japan's Secret War: Japan's Race Against Time to Build Its Own Atomic Bomb*, updated ed. (New York: Marlowe & Company, 1995); D. Oh, *Nishina Yoshio and Modern Physics in Japan*, 228–240; Kenji Ito, "Values of 'Pure Science': Nishina Yoshio's Wartime Discourse between Nationalism and Physics, 1940–1945," *Historical Studies in Physical and Biological Science, 33* (2002), 61–86 on 80–84. Keiko Nagase-Reimer, Walter E. Grunden, and Masakatsu Yamazaki, "Nuclear Weapons Research in Japan during the Second World War," *Historia Scientiarumm, 14* (2005), 201–240; Walter E. Grunden, *Secret Weapons and World War II: Japan in the Shadow of Big Science* (Lawrence, KS: University Press of Kansas, 2005); Morris Low, *Science and the Building of a New Japan*, (New York: Palgrave Macmillan, 2005), Chapter 2: Mobilizing Science in World War II: Yoshio Nishina.
83 For the English translation of memoranda between Nishina and Nobuuji, see MP4204 (American Institute of Physics). The short summary can be found in Richard Rhodes, *The Making of the Atomic Bomb* (New York: Touchstone, 1986), pp. 580–582.
84 Y. Nishina, "The Reconstruction of the Greater East Asian Co-Prosperity Sphere and Pure Science," *Kagaku, 12* (March, 1942), 1. For more information about Nishina's wartime activities, see Ito, "Values of 'Pure Science'," pp. 61–86.
85 *Report on Scientific Intelligence survey in Japan, September and October 1945*, National Archives, Record Group (RG) 165, Box 2299, 7/13/44/3, Vol. 1, 17; Vol. 3, appendix 3-A, 1. The short summary of Karl T. Compton's appraisal of the Japanese nuclear bomb project can be found in Rod W. Home and Morris F. Low, "Postwar Scientific Intelligence Missions to Japan," *Isis, 84* (1993), 527–537 on 533–534. For Compton's report, see also Yukuo Sasamato, "The Scientific Intelligence Survey: the Compton Survey," in Shigeru Nakayama (ed.), *A Social History of Science and Technology in Contemporary Japan. Vol. 1: The Occupation Period, 1945–1952* (Melbourne: Trans Pacific Press, 2001), pp. 59–72.
86 Major Russell A. Fisher, "Inspection of Activities at Kyoto Imperial University (February 28, 1946)," in *GHQ/SCAP Top Secret Records* (Tokyo: Kashiwashobo, 1998), Vol. 3, pp. 63–65 on p. 64.
87 Nagase-Reimer, Grunden, and Yamazaki, "Nuclear Weapons Research in Japan during the Second World War," 219. See also Grunden, *Secret Weapons & World War II*, Chapter 2: Nuclear Energy and the Atomic Bomb.
88 Walter E. Grunden, *Secret Weapons & World War II*, Chapter 3: Electric Weapons: Radar and the "Death Ray."
89 E. Tajima, "Remembering Nishina Sensei [in Japanese]," *Nihon Genshiryoku Gakkaishi, 32* (1990), 1167.
90 Although all cyclotrons were destroyed as ordered, several Van de Graff and Cockroft type accelerators survived. For the destruction of Japanese cyclotrons in 1945, see Shigeru Nakayama, "Destruction of Cyclotrons," in Nakayama (ed.), *A Social History of Science*

and Technology in Contemporary Japan, pp. 108–118. Some original memo and letters related to the incident were recently published. See *GHQ/SCAP Top Secret Records*, Vol. 6, folder (2): Cyclotron (file #13).

91 "Cyclotron Smashing: American Soldiers Demolish and Sink Precious Jap Scientific Equipment," *Life*, *19:26* (December 24, 1945), 26–27.

92 Thomas C. Smith, "The Kyoto Cyclotron," *Historia Scientiarum*, *12* (2002), 74–82 on 77–78.

93 For the twentieth century big science, see Peter Galison and Bruce Hevly (eds.), *Big Science: The Growth of Large-Scale Research* (Stanford: Stanford University Press, 1992).

94 Yoichiro Nambu interview with Dong-Won Kim (May 13–14, 1999).

95 Tajima, "The Cyclotron at Riken," 121–122.

96 Hinokawa, "Cyclotron Development at the IPCR," 32.

97 For the researchers at Lawrence's laboratory and their activities, see Heilbron and Seidel, *Lawrence and His Laboratory*, Chapters 4–6.

98 Y. Nishina to G. Hevesy (August 28, 1937) in *G. Hevesy–Y. Nishina Correspondence, 1928–1949*, pp. 19–20 on p. 20. Italics are added. Even though Nishina added that "I know that it is not an easy matter to construct and operate such a cyclotron smoothly," overall tone of the letter illustrates his confidence for the task.

99 Tomonaga, "Dr. Nishina," in Matsui and Ezawa (eds.), *Sin-itiro Tomonaga*, pp. 112–113.

100 See Heilbron and Seidel, *Lawrence and His Laboratory*, Table 6.5: U.S. cyclotrons by size, 1940 (p. 310) and Table 7.1: Foreign cyclotrons by size, 1940 (p. 321). The latter table shows that a similar sized small cyclotron (3 to 7 MeV) in Leningrad produced the first beam in September of 1937. Larger, medium-sized (8 to 12 MeV) cyclotrons in Cambridge (August, 1938), Copenhagen (November, 1938), Liverpool (mid-1939), Osaka (1939), Paris (March, 1939), and Stockholm (August, 1939) soon followed them.

101 Heilbron and Seidel, *Lawrence and His Laboratory,* Chapter 7: Technology transfer (especially Table 7.1).

102 For the Japanese reverse engineering in the early twentieth century, see Morris-Suzuki, *The Technological Transformation of Japan*, pp. 115–116 and 151–153.

7 Statesman of Science

When I would talk with anybody like Nishina, I would trust him like my brother.

Harry C. Kelly[1]

[Nishina] is doing a superb job both for the economic recovery of Japan and the development of international understanding.

Harry C. Kelly to Niels Bohr[2]

The present objective of our Institute [Kaken] is the application of science to peaceful industry in order to promote the rehabilitation of general economy of this country, in which the poverty paralyses its whole machinery. Scientists must take their due share in realizing the economical [*sic*] recovery of Japan in order that she can assume her responsibility in promoting world peace.

Yoshio Nishina to Niels Bohr[3]

On August 15, 1945, Emperor Hirohito announced that Japan would surrender unconditionally to the Allied Powers. The Japanese Empire's dream of a "Greater East Asia Co-Prosperity Sphere" had ended. Nine days after the first nuclear bomb had been dropped on Hiroshima and six days after a second such bomb had been dropped on Nagasaki, Hirohito addressed the nation:

> The enemy has begun to employ a new and most cruel bomb, the power of which to do damage is indeed incalculable, taking the toll of many innocent lives. Should we continue to fight, it would result not only in an ultimate collapse and obliteration of the Japanese nation, but also it would lead to the total extinction of human civilization. Such being the case, how are we to save the millions of our subjects, or to atone ourselves before the hallowed spirits of our imperial ancestors? This is the reason why we have ordered the acceptance of the provisions of the Joint [Potsdam] Declaration of the [Allied] Powers.[4]

The total defeat brought significant changes to the lives of all Japanese. Between the end of August 1945, when the U.S. forces landed in Japan, and April 28, 1952 when the San Francisco Peace Treaty took effect, Japan was occupied and governed by the Allied Powers that predominantly consisted of U.S. military. The Supreme Commander of the Allied Powers (SCAP), General Douglas MacArthur, ruled the nation from his General Headquarters (GHQ) in Tokyo. "Demilitarization" and "democratization" were the two basic principles of American policy in postwar Japan. Japan's army and navy were demobilized and equipment was seized or destroyed. Powerful government agencies like the Technology Agency were dissolved as were giant zaibatsu (conglomerates) like Mitsui or Mitsubishi. All scientific research with any military potential was banned. The destruction of the Japanese cyclotrons in November 1945 was an example of the U.S. demilitarization policy of Japan. After completing the "dismantling the Meiji State," new and more democratic systems replaced the old ones under the tutelage of GHQ.[5] Science and technology were no exceptions.

For Nishina, the end of the war meant the end of his research and teaching career. Initially he was naïve enough to believe that the Americans would allow him to continue his nuclear research with his large cyclotron. In his letter to GHQ on October 15, 1945, Nishina requested permission to operate his cyclotron at Riken for the "production and examination of radioactive substances as well as neutrons, both of which will be used in biology, medicine, chemistry, and metallurgy."[6] A few days later GHQ allowed him to "conduct investigations of radioactive substances and neutrons only in the fields of biology and medicine, and not in the fields of chemistry and metallurgy."[7] The destruction of cyclotrons in Riken along with those in Osaka and Kyoto in November, however, dashed his hope for further research. After that all nuclear experimental research was prohibited by GHQ. It finished Nishina's career as researcher, because he had concentrated all his efforts in the construction of cyclotron and in nuclear physics since the late 1930s. But, Yukawa and Tomonaga, both theoreticians, could continue their research. In the spring of 1946 Tomonaga started his famous "Tomonaga Seminar (or Friday Seminar)" in a classroom of Tokyo Bunrika University that attracted not only Tomonaga's own pupils at the Bunrika University but also students from the Department of Physics at Tokyo Imperial University. Tomonaga and young theoreticians elaborated his "super-many-time theory (renormalization theory)" amidst the ruins of Tokyo.[8] A few months later, in July 1946, Yukawa started a new journal in English, *Progress of Theoretical Physics*, devoted to original theoretical research.

Japan's desperate postwar condition, however, did not leave Nishina idle but bestowed upon him a new but equally important role — that of statesman of Japanese science. Pragmatist that he was, Nishina carried out this new role impressively. Nishina was the most influential senior liaison between the Japanese science community and GHQ for the reconstruction of Japanese science and technology. He was well suited for this mission. First, Nishina was internationally well-known as coauthor of Klein–Nishina formula, and had many distinguished friends in the West, including several Nobel Prize winners. Second, because of the nuclear attacks on Hiroshima and Nagasaki and because of the wartime Japanese nuclear bomb project, the United States sent physicists to survey Japanese science and technology, and they certainly knew of Nishina before coming to Japan. Third, Nishina could speak English well enough to communicate with Americans. Owing to these qualities, as Morris Low notes, "[Nishina] enjoyed a special rapport with the authorities, which placed him in a good position to negotiate the future of [Japan's] science."[9]

Nishina's reputation among American scientists was apparent from the very beginning. Karl T. Compton, physicist and the president of MIT since 1930, led the earliest scientific intelligence mission to Japan in September 1945.[10] He certainly knew Nishina through his scientific works. After interviewing Nishina at

Riken in September 18, 1945, Compton concluded his report with the following recommendation:

> It is clear that the present equipment, as well as the present program of Dr. Nishina, are suitable only for purely scientific work and could not be used for any military preparation except in the very long range way in which any scientific advance may potentially be a basis for a war application. Dr. Nishina is an excellent nuclear physicist who has previously cooperated internationally. In view of the above circumstances, I would recommend that the U.S. authorities permit him to go ahead with his scientific work, but that they prohibit him or any other Japanese scientists from undertaking the separation of uranium-235.[11]

Another American physicist, Harry Charles Kelly, likewise treated Nishina with respect.[12] Kelly, who held an MIT Ph.D., had worked on the electronic charge distribution in Townsend discharges with Philip Morse and John Slater. During World War II, William Shockley and Slater had recruited him to work on radar at MIT's Radiation Laboratory. In the winter of 1945, the Army sent him to Japan for scientific intelligence mission because he had "no connection with the Manhattan Project" so that "nobody can learn any secret from [him]."[13] Kelly and Gerald Fox, another physicist from the Radiation Laboratory, arrived in Tokyo in January 1946 and were assigned to the Scientific and Technical Division in the Economic and Scientific Section at GHQ. Kelly's job was to "keep surveillance over research and development in Japan, [and] advise GHQ on policy and action to be taken in regard to all matters pertaining to science and the teaching of science."[14] After Fox returned to the United States within a few months, Kelly assumed sole responsibility for the reorganization of science and technology in Japan until February 1950 when he and his family finally returned to the United States. He quickly earned the respect and cooperation from both GHQ and Japanese scientists and engineers, the latter of whom were first surprised by his frankness and then deeply impressed by his enthusiasm.

For Kelly, Nishina's importance was clear from the outset. In a 1975 interview with Charles Weiner, he recalled:

> I had great respect for some of the work of the Japanese that I knew. Nishina's work, for instance. Everybody knew about the Klein-Nishina formula and so on.... I guess he was the leader of all work in nuclear activity in Japan, and he was deeply respected by everyone there. So I would say, if you pick up one guy, he'd be it.[15]

Nishina became deeply involved in Kelly's reconstruction of science in post-war Japan.[16] In June 1946 Kelly organized the Japan Association of Science Liaison (JASL) as an advisory group to bridge GHQ with the Japanese scientific community. Nishina was one of 40 representatives from major universities and institutes. Kelly's ultimate goal was to create a national body for Japanese science that would be represented by democratically elected members. The first U.S. Scientific Advisory Group, which the U.S. National Academy of Science sent to Japan in the summer of 1947, endorsed that goal.[17] In July 1948, the Diet passed the law establishing the Science Council of Japan as the national scientific organization that would absorb prewar institutions like the Academy of Science in Japan. The 210 members of the seven divisions

FIGURE 7.1 Harry C. Kelly (second from the right) shaking hands with Naoto Kameyama, the first president of the Science Council of Japan. Kelly, the most important American for the reconstruction of the Japanese science and technology after World War II, regarded Nishina (far right) as the key partner. (Courtesy of the Special Collections, NCSU Libraries.)

of the Science Council were elected by the nation-wide vote of scientists and engineers on December 20, 1948. Nishina was not only elected as a member but also was appointed as one of the vice-presidents at the inaugural meeting on January 20, 1949 (Figure 7.1). At the same time, in January 1949, the Japanese government created the Scientific and Technical Administration Commission (STAC) in the prime minister's office as liaison between the government and the Science Council. Nishina was appointed acting director of the STAC in March 1949.

Nishina's most important postwar contribution to Japanese scientific community was the reorganization and revival of Riken.[18] Here Kelly's support was indispensable. Nishina wrote Lawrence that "Dr. Kelly has been doing a strenuous effort for the rehabilitation of science and technology in this country and I must say that we owe him the very existence of our Institute."[19] After the war, Riken had many serious difficulties and its survival was in question. Most of its buildings and facilities had been destroyed by American carpet bombings during the last days of the war, and it was heavily in debt. Okochi, who had successfully managed the institute for more than two decades, was arrested as suspected war criminal and imprisoned for four months. GHQ considered Riken (with its venture company Rikagaku Kyogo) as a zaibatsu and issued an order in June 1946 to dissolve it according to the antitrust law. Kelly,

FIGURE 7.2 After the end of the war, Nishina became the president of Riken and struggled for the survival of the institute. One of his solutions to overcome the financial difficulties was to manufacture penicillin using Riken's facilities. (Courtesy of the Special Collections, NCSU Libraries.)

however, was sympathetic to the cause of Riken, and tried to save it. Nishina was Kelly's principal contact in Riken during its reorganization.

In July 1946, Nishina was elected as one of 19 scientists to lead Riken. When Okochi and his board of directors formally resigned on October 25, 1946, Nishina was elected as Riken's fourth president (Figure 7.2). His immediate task was to continue Riken while he and Kelly were converting the institute from a foundation to a company, following GHQ's recommendation. Since Riken was in the midst of a financial crisis, Nishina had to find a way to avoid bankruptcy. He had to reduce the size of staff. He did so by firing old guards, including the legendary Kotaro Honda and even his former patron Nagaoka.[20] He also visited banks and GHQ to secure loans and Kelly helped him persuade GHQ. A more scientific and productive solution to the institute's financial problems was to manufacture penicillin in Riken.[21] Nishina decided to carry out the penicillin project because of the high domestic demand for the medicine, the availability of expertise in Riken (e.g. the high vacuum technology that he had developed for the cyclotrons), and the symbolic significance of penicillin as a peaceful application of science.[22] The research for the production of penicillin started in the fall of 1946 and Riken began to manufacture penicillin in 1947. In his letter to Lawrence in January 1947, Nishina described his new job as follows:

> Last November I was elected to the Presidency of our Institute, more than half of which was destroyed by air-raids. I became a sort of businessman for the time being, because

> there are tremendous work [*sic*] to be done for rehabilitation of our Institute, which is to become a new company instead of a juridical person as it has been for about 30 years. The first work I have to do is the repair of some buildings in order that we can start enough work for sustaining ourselves economically. For this I must get a loan from a bank and this is outside the domain of science. My objective is to reorganize our Institute in such a way that we can conduct a wide range of work, from fundamental scientific research to its application to industry, agriculture and medicine. This is not a small work, especially in such a difficult time as we experience now. I think, however, this is only way we can survive and at the same time can contribute to the rehabilitation of Japan, to the progress of science and thus to the benefit of the world. Now I am laying just the foundation of this work, which will take many years to come, and I have to put aside physics for the present.[23]

Nishina, who had been the champion of pure science until the end of the war, was also a pragmatist. Now, after the disastrous war that Japan had waged, he emphasized the importance of science for the rehabilitation of Japanese economy. He repeated the point in several articles that he wrote for newspapers and popular science journals. For example, in his article in *Shizen* (1946), he claimed that "as scientists and engineers it is our duty to make earnest efforts to revitalize industry."[24] This was what Kelly had repeatedly urged Japanese scientists to do. Samuel K. Coleman, in his article on the reorganization of Riken, summarizes Kelly's influence on Nishina as follows:

> Nishina was deeply moved by Kelly's insistence that scientists help with economic reconstruction. Kelly himself later credited Nishina with striving "to help his people recognize how the great advances in science and technology could help alleviate the grave problems of food, clothing, shelter and health." Tomonaga Shin-itiro recalled Nishina's request to pass along to his younger colleagues the ESS/ST scientists' exhortations to drop fundamental research and concentrate instead on applied problems.[25]

Old Riken was finally dissolved in February 29, 1948. The new company, Scientific Research Institute, Ltd. (Kagaku Kenkyusho, or Kaken), took over personnel, land, building, and equipment from the old Riken and was formally inaugurated on March 1, 1948. Nishina was elected as the first president. The 1948 brochure announced that the purpose and aim of the company were:

> the Advancement of Science and Technology and their Application to Industry for the Rehabilitation of [the] General Economy of Japan. The New Company will be available to Industrial Concern for Consultation on Technical Problems.[26]

While Nishina repeatedly emphasized Japanese scientists' responsibility for economic rehabilitation, he also reiterated the value of the peaceful use of science and nuclear power and of the international cooperation among scientists. *Atomic Power and I* (1950), the collection of articles that Nishina wrote for popular journals or newspapers after 1945, shows his pragmatic attitude on these issues. As Japan's postwar statesman of science, he tried to persuade his fellow Japanese to accept these three points that fitted well with GHQ's official positions. The articles in *Atomic Power and I* however were quite different from those articles that Nishina had written during

World War II, when he supported the Japanese imperial cause and emphasized the role of science for Japan's victory. The cause of this drastic change in his image, from the head of Japanese nuclear bomb project to the preacher of peaceful use of science or nuclear power, has been controversial among historians. Perhaps the following excerpt (ca. 1948) indicates his true stance — cautious but pragmatic.

> Scientists are not responsible for starting war. But when war is once started they are called upon to cooperate in its prosecution, and through their cooperation war is made increasingly terrible.
>
> If all scientists in the world unanimously decide not to cooperate in any war, it may go a long way toward preventing war, but such a proposition appears to be well-nigh impossible, as it is supremely difficult in existing circumstances for American scientists to come in contact and act in close concert with Soviet scientist in the matter. It is therefore doubtful whether the well-meant efforts of scientists to prevent war will prove successful. This consideration should not, however, deter them from putting forth earnest endeavors to promote peace.
>
> Scientists in America are collectively carrying on a peace movement. Dr. Einstein is a prominent figure in this movement. Although many American scientists are favorably inclined toward this movement, there are a few scientists who evince little interest in it. Their indifference is due not so much to their positive support of war as to their belief that in the last resort war is unavoidable.[27]

Nevertheless, Nishina's repeated emphasis on these important issues in the postwar period certainly helped the Japanese scientific community to overcome the militaristic image of Japanese science before 1945 and to replace it with a commitment to peaceful and useful application, thus preparing Japan's return to the international community.

The international political situation quickly turned in Japan's favor in 1948. The advent of the Cold War dramatically changed American policy toward Japan. Japan, once the reviled enemy, now became the United States' most important bulwark against communism in East Asia, and many of the strict restrictions that had been imposed on Japan's scientific efforts were relaxed. GHQ, which had denied Japanese scientists' request to travel abroad until the end of 1947, now allowed them to do so. The eminent geneticist, Hitoshi Kihara, left Japan in July 1948 to attend the Eighth International Conference of Genetics and became "the first Japanese scientist to travel overseas after WWII."[28] Yukawa became the second. Oppenheimer, then the director of the Institute of Advanced Study at Princeton, sent a letter to General MacArthur inviting Yukawa to the institute. GHQ granted permission and Yukawa left for the United States on September 2, 1948. He heard the news that he had won the 1949 Nobel Prize in physics while living in New York as a visiting professor at Columbia University. Tomonaga and many other Japanese physicists soon visited the United States. One statistics indicates that between the summer of 1948 and the end of 1951, 807 of 897 Japanese scientists who traveled or studied abroad chose the United States as their destination.[29]

Nishina also played a crucial role in Japan's return to the international science community. When the second U.S. Scientific Advisory Group visited Japan in late November 1948, among the guests was one of Nishina's old friends, Isidor I. Rabi

FIGURE 7.3 Nishna greeting Isidor I. Rabi, who came to Japan in November of 1948 as a member of the second U.S. Scientific Advisory Group. They had worked together under W. Pauli in Hamburg in the winter of 1927–1928 and coauthored a paper. Nishina played an important role of Japan's returning to the international science community. (Courtesy of AIP Emilio Segrè Visual Archive.)

(Figure 7.3). Rabi, professor of physics at Columbia University and the 1944 Nobel Prize winner in physics, had coauthored a paper with Nishina when they were in Hamburg in the winter of 1927–1928. He and other members of the Advisory Group admired the high quality of Japanese scientists and encouraged GHQ to send more young Japanese scientists overseas for advanced study and to cultivate the international exchange of senior Japanese scientists.[30] Rabi was especially enthusiastic in supporting Japanese scientists and recommended that the United States use them for the mutual benefit:

> Japan, in a number of fields, has scientists who are now ineffective because of the lack of equipment. In the United States, on the other hand, we have ample equipment and support but suffer from the shortage of scientific personnel. There are many cases where Japanese scientists would be very useful in researches in the United States, which are supported from military funds given to universities. In many cases these funds are inefficiently used because of the shortage of highly trained personnel in the United States.[31]

In September 1949, GHQ allowed Nishina to go to Copenhagen to attend the general meeting of the International Council of Scientific Unions (ICSU) as the representative of the Science Council of Japan.[32] He thereafter served as the official

Japanese delegate to the General Assembly of UNESCO in Paris. These two meetings signaled Japan's return to the international community. Letters between Bohr and Detlev W. Bronk (Foreign Secretary of the U.S. National Academy of Science and a member of the second Scientific Advisory Group in 1948), Kelly and Lawrence after Nishina's visit to Copenhagen suggest that the international science community was ready to accept Japanese scientists.[33] Lawrence and other American physicists were especially eager to invite Nishina to the United States.

In the spring of 1950, Naoto Kameyama (president), Sakae Wagatsuma (vice-president), and Nishina (vice-president) of the Science Council visited the United States. The National Leaders Project, which GHQ had established in 1949, and the U.S. National Academy of Science sponsored the visit.[34] They spent more than a month touring the National Academy of Science in Washington, DC, major universities and research institutes. During this visit, Nishina, who had corresponded with Lawrence in connection with the construction of Riken's cyclotrons since 1931, finally met him. In his postcard to Lawrence on the way back home, Nishina wrote, "It is just 9 hours' flight and I am still thinking of the most wonderful reception you gave me. The days spent with you are unforgettable."[35] Neither Nishina nor Lawrence suspected that this would be their last meeting.

Nishina remained dedicated to the development of new Riken (Kaken) and to the Japanese scientific community until his death in 1951, but the darkest days were certainly over. Penicillin brought Riken a steady income and the financial condition of the institute greatly improved. Research conditions also improved. In 1948, Nishina received permission from GHQ to use radioactive-isotopes for medical and industrial use. On November 10, 1949, the U.S. Atomic Energy Commission announced

FIGURE 7.4 Nishina opening the box that contained the radioactive isotopes in April 1950. (Courtesy of the Special Collections, NCSU Libraries.)

FIGURE 7.5 Nishna's tomb in Tama Reien Cemetery in Fuchu City, a western suburb of Tokyo. The letters in the tomb stone were written by Prime Minister Shigeru Yoshida. The obelisk shape tomb in the middle is Tomonaga's tomb. (Photo taken by the author.)

that Japan could import radioactive-isotopes for medical and scientific research, and STAC set up the Japan Radio-isotope Trade Association for that purpose in December. The American Philosophical Society subsidized the purchase of isotopes, and these were given to "the Nishina Laboratory at Kaken (Scientific Research Institute)." This would be Kelly's last present to Nishina. Delighted that Japanese scientists could finally resume their nuclear physics research, Nishina wrote an article, entitled "Radioactive-Isotopes Shall Be Imported," in the spring of 1950.[36] The first radioactive-isotopes arrived in Japan on April 10, 1950, and the second and third followed a month later (Figure 7.4). In July STAC authorized the Nishina Laboratory to distribute radioactive-isotopes in Japan.

As Nishina prepared to resume his research on nuclear physics, he suddenly became seriously ill and was diagnosed with liver cancer. Nishina's colleagues and former students had long been preparing for his 60th birthday party that was regarded as a great celebration for all East Asians. But it was cancelled in the last minutes. Nishina was hospitalized on December 12, six days after his birthday, and passed away on January 10, 1951. He was buried in Tama Reien Cemetery in Fuchu City, a western suburb of Tokyo (Figure 7.5). Twenty-eight years later, in 1979, Tomonaga, Nishina's faithful disciple, was buried by his side.

NOTES

1 Interview with Harry C. Kelly by Charles Weiner (October 27, 1975), Institute Archives and Special Collections, MIT Libraries, Cambridge, MA, pp. 77–117 and p. 46.

2 H. C. Kelly to N. Bohr (October 19, 1949), in *Supplement to the Publications*, pp. 35–36.
3 Y. Nishina to N. Bohr (August 17, 1948), in *Y. Nishina's Correspondence with Bohr and Copenhageners, 1928–1949*, pp. 62–63.
4 Robert J. C. Butow, *Japan's Decision to Surrender* (Stanford: Stanford University Press, 1954), p. 248.
5 Herbert Passin (ed.), *Remaking Japan: The American Occupation as New Deal* (New York: Free Press, 1987); Meirion Harries and Susie Harries, *Sheathing the Sword: The Demilitarisation of Japan* (London: Hamish Hamilton, 1987); Marius B. Jansen, *The Making of Modern Japan* (Cambridge, MA: Harvard University Press, 2000), pp. 666–690.
6 Y. Nishina to SCAP (October 15, 1945) in *GHQ/SCAP Top Secret Records*, Vol. 6, pp. 125–127.
7 GHQ to Imperial Japanese Government (October 27, 1945) in *GHQ/SCAP Top Secret Records*, Vol. 6, p. 117.
8 Matsui and Ezawa (eds.), *Sin-itiro Tomonaga*, Chapter 4: Amidst the Ruins, a Mecca for the Theory of Elementary Particles.
9 M. Low, *Science and the Building of a New Japan*, (New York: Palgrave Macmillan, 2005), p. 62.
10 *Report on Scientific Intelligence Survey in Japan, September and October, 1945, Vols. 1–5*, National Archives, Washington, DC, Record Group (RG) 165, Boxes 2055–2056, 390/33/21/2. See also Rod W. Home and Morris F. Low, "Postwar Scientific Intelligence Missions to Japan," *Isis, 84* (1993), 527–537; Shigeru Nakayama, "The Role of Advisory Missions," in S. Nakayama (ed.), *A Social History of Science and Technology in Contemporary Japan*, (Melbourne: Trans Pacific Press, 2001), (Melbourne: Trans Pacific Press, 2001), pp. 179–192.
11 *Report on Scientific Intelligence Survey in Japan*, Vol. 3, p. 6-a-3.
12 For Harry C. Kelly's activities in postwar Japan, see Samuel K. Coleman, "Riken from 1945 to 1948: The Reorganization of Japan's Physical and Chemical Research Institute under the American Occupation," *Technology and Culture, 31* (1990), 228–250; Hideo Yoshikawa and Joanne Kauffman, *Science Has No National Borders: Harry C. Kelly and the Reconstruction of Science and Technology in Postwar Japan* (Cambridge, MA: MIT Press, 1994); Interview with Harry C. Kelly by Charles Weiner (October 27, 1975); and M. Low, *Science and the Building of a New Japan*, Chapter 3.
13 Interview with Harry C. Kelly by Charles Weiner (October 27, 1975), p. 20.
14 Yoshikawa and Kauffman, *Science Has No National Borders*, p. 22.
15 Interview with Harry C. Kelly, pp. 42 and 49.
16 For the reconstruction of science in postwar Japan, Shigeru Nakayama, *Science, Technology and Society in Postwar Japan* (London: Kegan Paul International, 1991), Chapter 2; Nakayama (ed.), *A Social History of Science and Technology in Contemporary Japan*; M. Low, *Science and the Building of a New Japan*, Chapter 3.
17 Short summary of the activities of the first and second Scientific Advisory Groups are given in Shigeru Nakayama, "The Role of Advisory Missions," in Nakayama (ed.), *A Social History of Science and Technology in Contemporary Japan*, pp. 179–192 on pp. 181–184.
18 For the reorganization of Riken, see Coleman, "Riken from 1945 to 1948"; Yoshikawa and Kauffman, *Science Has No National Borders*, 84–91; H. Tamaki, "The Agonies of the Birth of Kaken [in Japanese]," *Shizen* (December, 1978), 153–159.
19 Y. Nishina to E. O. Lawrence (May 29, 1947), EOL 9/38.
20 Coleman, "Riken from 1945 to 1948," 244.
21 Shimpei Miyata, *Heaven of Freedom for Scientists: The Illustrations of the Institute of Physical and Chemical Research* [in Japanese] (Tokyo: Bungeishunju, 1983), pp. 274–276.

22 Coleman, "Riken from 1945 to 1948," 245–246.

23 Y. Nishina to E. O. Lawrence (January 3, 1947), EOL 9/38.

24 Y. Nishina, "Science, Technology and the Improvement of the Character of the People [in Japanese]," dated October 20, 1946, reprinted in *Shizen* (March, 1971), 19.

25 Coleman, "Riken from 1945 to 1948," 239.

26 This separate brochure is included in the letter from Y. Nishina to E. O. Lawrence (August 16, 1948), EOL 9/38.

27 English translation of excerpt from Y. Nishina's article in *Minron* (ca. 1948) in *GHQ/SCAP Records*, ESS (E)-06392, pp. 5–6.

28 S. Nakayama, "Sending Scientists Overseas," in Nakayama (ed.), *A Social History of Science and Technology in Contemporary Japan*, pp. 249–260 on p. 250.

29 Ibid., p. 252.

30 Low, *Science and the Building of a New Japan*, p. 52.

31 Ibid., p. 54.

32 For his trip to Copenhagen and Paris, see letters from Y. Nishina to N. Bohr (August 19, 1949; August 26, 1949; October 8, 1949) in *Y. Nishina's Correspondence with N. Bohr and Copenhageners, 1928–1949*, pp. 65–70.

33 Letters between Bohr and Bronk, Lawrence and Kelly, *Supplement to the Publications*, pp. 30–36.

34 Nakayama, "Sending Scientists Overseas," pp. 252–253.

35 Y. Nishina to E. O. Lawrence (April 4, 1950), EOL 9/38.

36 Y. Nishina, "Radioactive-Isotopes Shall be Imported [in Japanese]" in *Atomic Power and I*, pp. 121–126. The article was originally published in *Kagaku* (April, 1950).

Appendix

SELECTED CORRESPONDENCE BETWEEN YOSHIO NISHINA AND ERNEST O. LAWRENCE (1931–1951)

(*Source:* Ernest O. Lawrence Papers, The Bancroft Library, University of California, Berkeley, Carton 9, Folder 38 [Yoshio Nishina])

Yoshio Nishina and Ernest O. Lawrence shared many things in common: they loved engineering and paid more attention to the *construction* rather than *use* of the cyclotron; they knew how to utilize every possible resource for the construction of the cyclotron; and both were great bosses in managing a large number of young scientists under them. Although they had exchanged more than 100 letters since 1931, they met only once in 1950, about a half year before Nishina's sudden death. It was the cyclotron, and not common educational background or any experience under the same great scientist, that linked these two great scientists together. It is true that Nishina owed a great deal to Lawrence in the construction of his two cyclotrons. Lawrence, however, equally benefited from this relation and acquired a most dedicated and sincere *partner* outside the United States, who proved the superiority of his cyclotron models. Since their relation was so unique, it often provoked some misunderstanding, such as a description that Nishina was the passive client of Lawrence. The thick correspondence file in the Ernest O. Lawrence Archive (Carton 9, Folder 38) at Berkeley, however, indicates that their relation was deeper than that between giver and taker. The following are some selected letters that prove their unique partnership across the Pacific Ocean.

Lawrence to Nishina (December 5, 1931)

Dear Dr. Nishina:

I am enclosing herewith a copy of the abstract of a paper given before the National Academy on the production of high speed protons. Unfortunately we have not had time to write up a detailed account of our experiments, but we expect to do so some time during the next month. When this prospective manuscript is published, I shall remember to send you a reprint.

Nishina to Lawrence (April 21, 1932)

Dear Professor Lawrence,

I must thank you very much indeed for your kind letter of the 5th Dec. last and a separate copy of your interesting note in "Science" on the production of high speed proton. As I have been ill for some time, I have been prevented from writing to you to thank for your kindness. I am much interested in your work and should be very much obliged to you if you would send me a reprint, when you publish the detailed account of your experiments, for which I am waiting with a keen interest.

Nishina to Lawrence (July 30, 1936)

Through Sagane I have been following with much admiration the achievements which you have recently had with your cyclotrons. Our Cyclotron is now under construction with the help of experiences obtained by Yasaki in your laboratory as well as informations from Sagane. We hope to complete it in a few months' time.

The magnet was originally used for Poulsen arc generator and is of diameter 66 cm. Since the field intensity is rather low (about 14,000 gauss), the energy obtained is not high enough. We are consequently trying to get a fund for the construction of a larger magnet of diameter 100 or 125 cm. In this connection I should like to know the cost of such a magnet in America and should be much obliged to you, if you would let Sagane know the suitable manufacturer in your country. He will then get the necessary informations for me.

Lawrence to Nishina (August 18, 1936)

. . .

I have been talking to Dr. Sagane about magnet costs and together we will make every effort to find out the cheapest way for you to build a large magnet. Apparently the estimates you obtained from Japanese concerns are more than double the prices in this country, and therefore it might be most economical to order the iron pieces from America and have them shipped to Japan. . . .

We have enjoyed very much having Dr. Sagane with us this year. He has become a valued member of the laboratory, and we are very sorry that he is leaving, but understand perfectly well the desirability of spending the next year in the Cavendish Laboratory.

Nishina to Lawrence (January 8, 1937)

. . .

Now we have quite recently obtained a prospect of receiving the necessary fund. Yesterday I therefore asked you by cable the name of manufacturer and the cost of your 200-ton new magnet, of which Sagane wrote to me some time ago. If the cost is within our reach, I should like to have a similar one made for us. . . .

Our small cyclotron is near its completion. The building to house it has taken rather a long time but has just been finished and we hope soon to begin the operation. If we succeed in it, it is due to nothing but the kind information given us by you through Sagane and Yasaki.

Nishina to Lawrence (January 13, 1937)

Dear Professor Lawrence,

In order to express our gratitude to you for all the kindness you have shown to us in connexion with the operation and construction of our cyclotron, we send you today a cloisonné vase, which I hope will arrive to you at about the same time as this letter. There may be some delay due to the strikes in the harbour of San Francisco.

Nishina to Lawrence (January 20, 1937)

A few days ago we were informed by Mitsui Company here that your new magnet has not been ordered yet and the cost is not known at present. They told us that the estimates for 200-ton magnet will be obtained for us from some steel companies in America.

Now I should like to ask you whether it is possible for you to order for us the same magnet as yours when you order your own. . . .

There are some reasons why I am asking you to do us such a favour. Firstly we have not much experiences [sic] with such a large magnet and we have to depend on your advice. Secondly Mitsui people in New York have no knowledge of magnets and we shall lose much time in correspondence.

Lawrence to Nishina (February 18, 1937)

I would be more than glad to assist you in the purchase of an electromagnet and I had hoped before now to have information to cable you. Unfortunately, however, I have run into difficulties in negotiations for our own magnet and the order has been held up.

. . .

The thought occurs to me that in the event that you order the electro-magnet in this country this spring, it might be desirable for Dr. Sagane to return here somewhat sooner than anticipated, in order to take active charge of the negotiations for the purchase of the equipment and official supervision of the construction and all arrangements. I could perhaps be of more useful service to you if Dr. Sagane were here and I could assist directly by consulting and advising him.

Lawrence to Nishina (April 13, 1937)

Some time ago I wrote Dr. Sagane, suggesting that it might be well for him to return to California to take active charge of the negotiations for your magnet, and I have just received a letter from him saying that it was questionable in his mind whether he should do this. I have today dispatched a letter to him saying that now that the orders have been placed I see no reason at all why he should hurry back. It will be quite all right for him to arrive in July, when the steel perhaps will be here and the machining under way.

Nishina to Lawrence (June 1, 1937)

. . .

First of all I must tell you how much we appreciate your very kind and hard endeavour for purchasing our large electromagnet. Without your assistance it would have been simply impossible for us to obtain it for such a price. Through the Mitsui Company who has been supplying me with copies of all their correspondence concerned, I have been constantly informed of your hard fight with makers as regards the negotiations as well as of your all necessary supervision of the construction. We are all very thankful for your immeasurable assistance.

. . .

The blueprint of the magnet steel, which you kindly sent to me, arrived here some time ago. I quite agree with you that the ample space between the pole faces and outside yokes is very important for the convenient manipulation of the cyclotron. From the blueprint I see that the cross-sectional area of the horizontal yoke is larger than that of the vertical one. This was presumably done for preventing the bending of the horizontal yoke.

I should like to know the necessary power for exciting the magnet. As I informed you by cable some time ago, we should like preferably to use 500 volts for this purpose, although it is not absolutely necessary. It might be necessary to install a new D. C. generator for it as the case may demand, and I should be much obliged to you, if you would let me know the necessary voltage and current for this purpose.

I should like also to know the method of cooling of the magnet coils. If they are to be cooled by oil, oil tanks have to be made, and I should be much obliged to you, if you would let me know whether such tanks are included in our plan.

As Yasaki wrote to you some time ago, our cyclotron has been working since the beginning of April. Current is still small, 1 to 2 microamperes, which we are trying to increase. We have done some experiments on the effect of neutrons on the living organism and have obtained some interesting results. I hope to be able to write you about them in my next letter.

As I informed you by cable, Sagane is expected to return to Berkeley on about the 20th June. I wrote to him to supervise under your guidance the construction of the magnet steel, winding of coils, etc.

. . .

We should like very much to invite you and Mrs. Lawrence here to Japan some time next year or year after next, and I am contemplating how to realize it.

Lawrence to Nishina (July 2, 1937)

. . .

I am glad that you are interested in the biological effects produced by neutrons and shall look forward to hearing of your progress in this direction. There can be no question as to the importance of the biological work opened up by the availability of neutron rays and the artificial radioactive substances and I am glad that so many laboratories having cyclotrons recognize this and are carrying on biological investigations along with the nuclear work. Sagane arrived yesterday and already has gathered the information about the large magnet and cyclotron requested in your letter and I shall leave it to him to transmit it to you.

. . .

Of course I would have a particular pleasure in visiting your laboratory with the new cyclotrons in operation and with so much important work in progress. It is a question as to what time would be most suitable. I believe we could come over next spring or fall but perhaps it might be better to defer the trip for a year, when the new cyclotron should be completed and in operation.

Nishina to Lawrence (July 13, 1937)

. . .

In our laboratory our medical colleague, Professor Nakaidzumi and his co-workers have been studying the effects on the living cells of neutrons (accompanied by gamma-rays) produced by our cyclotron. I enclosed the copy of their note which has been sent to Nature. . . .

I presume that Sagane is already in your laboratory and is supervising under your direction the construction of the large magnet. . . .

Lawrence to Nishina (August 10, 1937)

At the moment we are installing our new 37 inch cyclotron chamber, and we hope to try it out within a fortnight. . . .

Nishina to Lawrence (February 21, 1938)

I am very sorry that I have not written to you a single letter since last summer. You have supplied us through Sagane with all necessary information regarding the new large cyclotron and I must tell you how much we have appreciated your immeasurable assistance and advices in this matter.

The magnet steel arrived here about the middle of January and is now in the Ishikawazima Dockyard for machining, which will probably take some months. We are looking forward to having it finished.

. . .

Sagane arrived here on Feb. 2nd and has supplied us with many useful informations [sic] in connection with the cyclotron as well as with other problems. Due to his shimming, for example, our cyclotron has increased its currents from about 15 μ.a. to nearly 30 μ.a. In this connection I heartily thank you for having Sagane in your laboratory for so many months. It is entirely due to the kind assistance and useful advices which you gave us through Sagane and Yasaki that we succeeded in the construction and the operation of our first cyclotron. We hope that the same will be the case with the second one, for which you have already given so much assistance.

With the first cyclotron we have been working on nuclear as well as biological problems. . . .

Nishina to Lawrence (May 2, 1938)

. . .

The machining of our magnet steel has nearly been finished and we have started erecting the magnet. All the coils and oil tanks have just arrived here and we are contemplating how to build them up. . . .

Nishina to Lawrence (February 26, 1940)

. . .

Last summer I learned the successful operation of your 60 inch cyclotron through your letter to Sagane. Also at that time I thought to write to you a letter of congratulation which has never been realized. I read in the *Science News Letters* since then that you are planning to build a cyclotron more than ten times heavier. Yukawa who visited you last October also brought me the same news. That would be a great thing and would contribute a good deal to the advancement of science as a whole. I am very much looking forward to its early realization.

As to our 60 inch cyclotron, it has taken a long time, and now a part has to be reconstructed. Sagane on his return advised us to make the gab of the magnet 22 inches instead of 20 inches as was originally planned. After following his advice, however, we found that the field strength drops too much at the edge. We tried to improve it by shimming but did not succeed and consequently we could not get the beam out of the dees of the cyclotron. We are now going to change the gab to 20 inches. That will take some time but I think everything is then in order.

Nishina to Lawrence (June 15, 1940)

During the coming summer vacation, two members of our laboratory, T. Yasaki and S. Watanabe will be sent to America to visit some nuclear physics laboratories. I should be very much obliged to you, if you would kindly allow them to stay in your laboratory for about a month.

...

I have been thinking of coming myself to Mecca of cyclotronists in order to have the pleasure of your personal acquaintance and also of learning various matters concerning the 60″ cyclotron, our edition of which is not in right operating conditions yet. My health, however, does not allow me to leave here at present and I let my assistants come to you instead.

You know Yasaki of course. Watanabe has done much in the design and construction of both of our cyclotrons. I expect that their knowledge and acquaintances obtained in your laboratory will be very valuable ones and serve much to strengthen the ties between two laboratories.

Lawrence to Nishina (August 22, 1940): telegram

ORDER JUST ISSUED BY UNIVERSITY PRESIDENT NOT PERMITTING VISITING SCIENTISTS IN LABORATORY. PLEASE NOTIFY YASAKI AND WATANABE WITH MY PERSONAL REGRETS.

Lawrence to Nishina (August 22, 1940)

...

The reasons for this sudden restriction have to do with the fact that in many of laboratories of the university there is at the present time marked overcrowding and also there is certain amount of work in progress of a confidential character.

...

I personally regret very much that this sudden order from the university authorities makes it impossible for me to welcome my old friend, Dr. Yasaki, and Dr. Watanabe to our laboratory. I can only hope now that they will be able to visit laboratories in other universities in this country and thereby make their trip worthwhile. . . .

Nishina to Lawrence (November 28, 1940)

A few days ago Yasaki, Watanabe and Iimori of our laboratory came home and told me of all the friendliness and generosity you and your colleagues have shown towards them. In view of the present international situation, I thank you from the bottom of my heart for what you have done for us in the cause of science and friendship.

...

I was told by Yasaki, Watanabe and Iimori what a great stride you have been doing in various fields of science by means of your 37″ and 60″ cyclotrons and we all admire your great success.

From what I have heard I have decided to reconstruct various parts of our 60″ cyclotron according to your line of construction. . . .

Nishina to Lawrence (July 15, 1946)

Dear Dr. Lawrence:

It is a long time since I last wrote to you. During that period a good deal has happened between us. A few days ago it happened to me to read your old kind letters and I realized what a great change we have experienced. The 60″ cyclotron which you were so kind to help us construct is unfortunately gone now forever deep in the Pacific. It may be said that it was constructed for being destroyed, because we could not use it for research very much on account of the war.

As Dr. Kelly, whose home journey I am availing myself for writing this letter, may tell you, our Institute will be changed into a new company, which we are going soon to organize and start afresh. I might work as an organizer for our new company and then have to drop physics for some years. . . .

P.S. We are very much interested in Dr. McMillan's Synchroton which I understand is soon to be built. How is the progress of the construction of the 5,000-ton cyclotron? I wish you a great success.

Lawrence to Nishina (October 31, 1946)

. . .

Knowing of your present difficulties in getting periodicals, I have arranged to have *PHYSICAL REVIEW, THE REVIEWS OF MODERN PHYSICS* and *THE REVIEW OF SCIENTIFIC INSTRUMENTS* sent to you for a year or longer, let us say until more normal arrangements of international exchange are achieved. . . .

Nishina to Lawrence (January 3, 1947)

. . .

It was very kind of you to have arranged to get Physical Review, The Reviews of Modern Physics and The Review of Scientific Instruments sent to me until "more normal arrangements of international exchange are achieved."

. . .

Here everything is still in the state of instability. Last November I was elected to the Presidency of our Institute, more than half of which was destroyed by air-raids. I became a sort of businessman for the time being, because there are tremendous work to be done for rehabilitation of our Institute, which is to become a new company instead of a juridical person as it has been for about 30 years. The first work I have to do is the repair of some buildings in order that we can start enough work for sustaining ourselves economically. For this I must get a loan from a bank and this is outside the domain of science. My objective is to reorganize our Institute in such a way that we can conduct a wide range of work, from fundamental scientific research to its application to industry, agriculture and medicine. This is not a small work, especially in such a difficult time as we experience now. I think, however, this is only way we can survive and at the same time can contribute to the rehabilitation of Japan, to the progress of science and thus to the benefit of the world. Now I am laying just the foundation of this work, which will take many years to come, and I have to put aside physics for the present.

Nishina to Lawrence (May 29, 1947)

. . .

First of all I must thank you for your kind thought in sending me [a] number of parcels containing beautiful candies and delicious dried fruits. I cannot be too thankful to you.

[list of items — cigarettes, candy bars, etc.]

. . .

Every time your parcel arrived here, my colleagues and assistants in our laboratory as well as my boys were simply delighted and enjoyed the taste of your precious gifts. You probably do not know how we value sweets, which are very scarce here. I have two boys, the elder being 17 years old and the younger 15. Since our house was burnt by air-raids during the war, we are living in one of the laboratory rooms of our Institute. Every time we take your sweets in the evening, I tell my boys about your work in which they have a great interest.

. . .

For last six months I have been engaged in the reorganization and rehabilitation of our Institute. The work does not proceed smoothly, but oscillates about an equilibrium point, which is making a very slow forward movement under internal and external influences. In this respect Dr. Kelly has been doing a strenuous effort for the rehabilitation of science and technology in this country and I must say that we owe him the very existence of our Institute. But for him it would have been impossible for us to keep our present organization. Now that we have just been placed on a right track, we can go ahead though slowly. My work now is to make the economy of our Institute self-supporting. This is not an easy task for a scientist but I have got to do it. . . .

Copy of Announcement

The president
and Board of Directors
of
The Scientific Research Institute, Limited
Desires to Announce
the Formation of the New Company on March 1, 1948. The
Personnel, Land, Buildings and Equipments of the Company
were taken over from the Former Institute of Physical and
Chemical Research, which was dissolved.

The Purposes and Aims of the New Company are the Advancement of Science and Technology and their Application to Industry for the Rehabilitation of General Economy of Japan. The New Company will be available to Industrial Concerns for Consultation on Technical Problems.

The Address of the New Company is:
31 Kamifujimae-cho, Komagome, Bunkyo-ku, Tokyo.

YOSHIO NISHINA, Sc. D.
President,
Scientific Research Institute, Ltd.

Nishina to Lawrence (August 16, 1948)

Yukawa told me that he will visit you on his way to Princeton and . . . You sent to me so many parcels which contained such nice things as we have never seen in our life.

. . .

The administration of the Institute is not an easy matter at this time of national hardship. The most difficult thing is the finance of the Institute, on which we have to concentrate our energy. I had to give up physics and become an administrator and a businessman. The present objective of our Institute is the application of science to peaceful industry and thus to promote the rehabilitation of general economy of this country, in which the poverty paralyze the whole machinery of the people. Scientists must take their due share in realizing the economical recovery of Japan in order that she can assume her responsibility in promoting world peace.

Lawrence to Nishina (January 7, 1949)

I have just received by mail from Professor Rabi the beautiful pearl necklace and Mrs. Lawrence and I don't know how to thank you for this wonderful gift. The pearls are beautiful beyond words and Mrs. Lawrence will always treasure them as a wonderful token of your friendship.

. . .

At the moment we are having an unusually interesting time here in the laboratory. We have just completed the modification of the 184 inch cyclotron for the production of 350 Mev protons and we are finding that these high energy protons produce considerable numbers of mesons as expected; also, by charge exchange we have a beam of 350 Mev neutrons. The 300 Mev synchrotron is now getting into operation also and accordingly, as you may well imagine, we are all very busy.

Nishina to Lawrence (February 20, 1950)

At the invitation of the United States National Academy of Science, the three officers of the Science Council of Japan are going to leave here on March 3 for Washington. They are Dr. Naoto Kameyama, the President, Dr. Sakae Wagatsuma and myself, the Vice-Presidents.

. . .

We shall take a PAA plane which arrive at 0100 on March 4 in San Francisco and leave at 2130 on the same day according to the time table. We, however, cannot expect the plane to be punctual and the error may be a matter of several hours according to my experience. Anyhow I shall get in touch with you by phone and if I have time I should like to come up to you. It has been my wish for nearly twenty years to see you and your laboratory and I am delighted at the thought that my wish will be fulfilled.

. . .

On my way back to Tokyo, I should like to visit Berkeley again and spend a day or two. I am much looking forward to seeing you in person, not in picture this time.

Lawrence to Nishina (March 9, 1950)

I was very sorry to miss your phone call when you passed through . . .

I hope that you will be able to spend some time with us here and, in that connection, as Don Cooksey also mentioned we would be glad to help with defraying your personal expenses in Berkeley.

I have known your name for so many years and more recently from our correspondence and have come to feel that I know you personally, and so I am looking forward ever so much to your visit.

Nishina to Lawrence (March 26, 1950)

Dear Dr. Lawrence:

I thank you very much indeed for your kind letter which I received at Washington. We are now visiting various research and educational institutions in this country and I expect to come to San Francisco at the beginning of April, possibly on April 2nd or 3rd and shall get in touch as soon as I arrive there.

I appreciate very much your suggestion to stay some time in Berkeley with your defraying my personal expense. It would be very pleasant and profitable for me to do so, but our Institute in Tokyo wants me to come home as soon as possible and to my great regret I cannot accept your invitation this time.

I am very much looking forward to seeing you.

Nishina to Lawrence (April 4, 1950): postcard

Dear Dr. Lawrence:

It is just 9 hours' flight and I am in Honolulu. I am still thinking of the most wonderful reception you gave me. The days spent with you are unforgettable.

Nishina to Lawrence (September 25, 1950)

I am very ashamed of not having written to you since I came home last April, although my thoughts always come to you, your family and your Laboratory whenever I recollect the last journey to the States. It was indeed a great pleasure and privilege for me to have seen yourself and your family for the first time, which I wished to do for so many years. To come to the Radiation Laboratory was also a great delight for me, because I had been acquainted with the name and had been admiring the great achievements from its beginning. As I told you at the time of my visit, your laboratory and people gave me the greatest impression I had had of recent years.

. . .

After coming home from the United States, I have entirely been taken up with the administration of my Institute and various Government and private committees. I may say I have no free time of my own. Since last June I have been working as an Acting Chairman for the Foreign Investment Commission, which deals with the introduction into Japanese industry of foreign technology as investment in kind. . . .

Some time ago the Tokyo radio broadcasting told us of your natural colour television, of which you told me at your home last March. We have not television here as yet. It would, however, be very interesting to introduce your method here in Japan, when the television is started in this country. . . .

P. S. The international situation has changed a great deal since I saw you last. We are very glad that a great advance is being made at the Korean war front. We must be very firm against communism.

Lawrence to Nishina (October 30, 1950)

. . .

As regard color television, nothing would give me greater pleasure than to have my color television inventions put into use in Japan. There is no doubt about it that color television is quite superior to black and white but how soon it will get into general use is of course uncertain.

The Korean war situation is surely now well in hand and let us hope that recent developments mark a real turning point in history.

Lawrence to Barkas (January 9, 1951)

I received a Christmas card from Nishina indicating that he is in a hospital seriously ill.

Naoto Kameyama (President of the Science Council of Japan) to Lawrence (January 12, 1951): telegram

INFORM DECEASE OF DR. YOSHIO NISHINA WITH HEARTFELT GRIEF.

Lawrence to Kameyama (January 13, 1951): telegram

WE ARE ALL SADDENED BY THE NEWS OF DR. NISHINA'S DEATH. HE WAS A WONDERFUL LEADER AND GREAT SCIENTIST WHOSE LOSS WILL BE FELT THE WORLD OVER. PLEASE CONVEY MY SINCERE SYMPATHY TO MRS. NISHINA AND FAMILY.

Kameyama to Lawrence (February 21, 1951)

...

The death of Dr. Nishina has been a heavy blow to his colleagues and to the scientific circles here. As time goes by and we recall his activities, the more keenly do we feel the loss of a great scientist, researcher, organizer, administrator and a sociable man, who had cultivated good relations with the scientific world across seas.

Fumio Yasaki, Asao Sugimoto, Hidehiko Tamaki and Eizo Tajima to Lawrence (October 24, 1951)

We are much obliged to you for your kind and instructive suggestions to the Japanese academic circle of physics at your visit here last summer. You impressed us very much by having encouraged the boys of the late Dr. Nishina and the staff of his laboratory.

Being stimulated by your visit, we have decided to reconstruct a small cyclotron, the size of which is same as the one we lost the other year. Fortunately we have a magnet which was originally used for a Poulsen arc generator. ...

Lawrence to Yasaki/Sugimoto/Tamaki/Tajima (November 13, 1951)

I am so glad to get your letter and to know that you are actually proceeding with the reconstruction of a small cyclotron. All of your friends over here are very pleased and hope that you will have every success in the undertaking.

Index